CAMBRIDGE COUNTY GEOGRAPHIES

General Editor: F. H. H. GUILLEMARD, M.A., M.D.

GLAMORGANSHIRE

T0345977

Cambridge County Geographies

GLAMORGANSHIRE

by

J. H. WADE, M.A.

Joint Author of *Rambles in Somerset* and the *Little Guides* to
Somerset, Monmouthshire, and South Wales.

With Maps, Diagrams and Illustrations

Cambridge:
at the University Press
1914

CAMBRIDGE UNIVERSITY PRESS
Cambridge, New York, Melbourne, Madrid, Cape Town,
Singapore, São Paulo, Delhi, Mexico City

Cambridge University Press
The Edinburgh Building, Cambridge CB2 8RU, UK

Published in the United States of America by Cambridge University Press, New York

www.cambridge.org
Information on this title: www.cambridge.org/9781107619722

First published 1914
First paperback edition 2013

A catalogue record for this publication is available from the British Library

ISBN 978-1-107-61972-2 Paperback

PREFACE

I SHOULD like gratefully to acknowledge the help which I have received in the compilation of this book from the Cardiff Naturalists' Society, who have supplied me with much information and have furnished me with some of the illustrations.

My thanks are also due to Professor G. W. Wade and Dr C. T. Vachell for kindly reading the proofs.

J. H. W.

March 1914.

CONTENTS

ILLUSTRATIONS

The illustrations on pages 19, 121, 123, 127, and 129 are from blocks kindly lent by the Cardiff Naturalists' Society. The Ogam Stone on p. 126 is sketched from an illustration in the *Archaeologia Cambrensis*. The Geological Section on p. 18 is from a map by Messrs G. Philip and Son.

The illustrations on pages 4, 7, 11, 15, 24, 34, 47, 50, 53, 54, 58, 62, 100, 102, 104, 112, 125, 135, 137, 139, 140, 142, 143, 144, 148, 149, 153, 156, 161, 173, 175, and 187, are reproduced from photographs by Messrs Frith & Co.; those on pp. 41, 76, 88, 93, 138, 147, and 163, are from photographs by Mr F. Evans; those on pp. 9, 14, 31, 84, and 164, are from photographs by Mr Osborne Long; those on pp. 61, 91, and 151, are reproduced by permission of the G. W. Rly. Co.; those on pp. 167 and 171 are from photographs by Mr Alfred Freke; that on p. 26 is from a photograph by Mr W. T. Cooper; and that on p. 183 is from a photograph by the Royal Photographic Co.   The sketch map of the castles is from a drawing by Mr C. J. Evans.

1. County and Shire. The name *Glamorgan.*

"County" and "shire" are now loosely used as equivalent terms. Originally they admitted of a distinction. "Shire" is an Anglo-Saxon word which at first denoted a portion of land "shorn" (for the words have the same derivation) from a larger territory for the satisfaction of a particular tribe. With the consolidation of the Anglo-Saxon rule the word lost its early tribal significance, and came to mean merely a territorial division for the administration of justice and for the collection of taxes. The "shire-reeve" (sheriff) was the official responsible for the discharge of both these functions. As Wales was never conquered by the Saxons, "shire" in the early sense of the word had no application to the Principality. Our Teutonic ancestors referred to its inaccessible fastnesses as "Wales," the land of strangers. The Welsh called it "Cymru," the land of fellow-country-men. It retained its ancient political independence until it was conquered piecemeal by the Normans; and when it was eventually annexed by the English crown and parcelled out after the English model into shires, the latter word had long acquired its purely administrative meaning.

Glamorganshire, though regarded from Norman times as a shire, had a much better claim to be called a county, for the latter term was of Norman introduction, and represented the Norman method of local government. Though the Normans retained the English form of political administration, they altered its character. Their method of government was less democratic and more arbitrary. The land passed by right of conquest from the people into the hands of the king, who let it out on feudal tenure to his counts, and each shire in consequence became a county. As the predominant feature of Norman rule was government by force, the natives were assessed in men as well as in money, and "county" and "shire" became the names of the same area in its military and civil aspects. The sheriff collected the king's revenue and administered the king's justice, and the count commanded the king's men. In the border counties the control of the Crown was much weaker, and in Wales, where dominion was gradually acquired by private adventure, it scarcely existed at all. The conquered territory was looked upon more or less as the personal property of the lord who secured it and he governed it much as he pleased. As Glamorganshire on its acquisition by the Normans obtained a regular administration of justice nominally under the jurisdiction of the crown, it was technically regarded as a shire, though it was really ruled by its lord, whose officer its sheriff was.

In the reign of Henry VIII, when the whole of Wales was formally incorporated with the English dominions and divided into shires, the boundaries of the original county of Glamorgan underwent a slight alteration. The limits of the Norman lordship had followed the lines of the old Welsh kingdom of Morganwg, and had extended from the Usk to the Tawe. Under the Tudor readjustment the district of Gwynllwg between the Usk and the Rhymney was thrown into the newly-formed shire of Monmouth, and by way of compensation the lordship of Gower was added to the county of Glamorgan.

Under the modern system of local government, "county" and "shire" are again beginning to lose their acquired identity. The "shire" is now little more than a geographical division, and the county has once more become the real administrative area. Their limits are no longer quite the same. From the county of Glamorgan have been taken away the county boroughs of Cardiff, Swansea, and Merthyr, which for governmental purposes are independent units, though they still belong geographically to Glamorganshire.

The name "Glamorgan" is merely a popular corruption of *Gwlad-Morgan*, "the land of Morgan," an appellation which it derived from one of its early princes. Its alternative title *Morganwg* arose from a common Welsh habit of designating a territory by adding *wg* to the name of its ruler.

2. General Characteristics.

The predominant feature of Glamorganshire is its commercial importance. It is not only the foremost county in Wales, but one of the richest provinces in the kingdom. Its industrial development has been one of

Three Cliffs Bay and Pennard Castle

the wonders of the age. A century and a half ago half of the shire was a highland wilderness valuable only to the sportsman and the sheep farmer. To-day these once solitary wastes are some of the most thickly populated districts in Britain. The physical features of the county have conspired to give it this leading commercial position. Like the rest of Wales it is in parts exceedingly mountainous. The hills lie piled up in great masses across half the county, but instead of barring its progress, they have been the chief cause of its phenomenal prosperity. Figuratively speaking, they have proved to be mountains of gold, for they are a vast storehouse of mineral treasure. Their yield of coal is prodigious, and there are immense deposits of limestone, as well as some iron ore.

Second only in importance to the mountains is the extensive sea-board with which the county is fringed. Though a great part of the coast is commercially useless, and the number of its natural harbours are comparatively few, yet it possesses several tidal estuaries which engineering skill has converted into docks, and the Glamorganshire ports are some of the busiest shipping centres in the kingdom.

But its rich mineral deposits are not the only source of wealth which the county possesses. Glamorgan was once famous for its fertility, and to-day it does not altogether belie its early agricultural reputation. Between the hills and the sea rolls a wide undulating plain which provides extensive pasturage for cattle, and furnishes an admirable soil for the cultivation of wheat. And the hills, though of little use in themselves for agricultural purposes, nevertheless form a serviceable screen for the crops in the lowlands.

Glamorganshire, besides being a wealthy and bountiful land, abounds also in historical interests. Few counties possess so many memorials of antiquity. Everywhere are to be found traces of primitive life, as well as of the civilisation which succeeded it. On the hills are the earthworks and sepulchral monuments of the disinherited Celts, and the plains are studded with the ruined castles of the invaders who supplanted them.

Artistically, too, the county is not without its attractions, though it has sacrificed much of its former beauty to its commercial prosperity. It no longer preserves the clear streams and wooded dells for which it was once famous, for the rivers are polluted and the valleys are sombre and sunless. The smoke of innumerable collieries and furnaces clouds the atmosphere, and the once luxuriant vegetation has been replaced by tiers of cottages; but outside the industrial districts many of its charms survive. The hills remain massive and majestic, and their rugged outlines and far-reaching prospects still charm the lover of scenery. The most fascinating region is the coast, which in places is quite remarkable for its grandeur. The Gower peninsula is in this respect especially notable. Its precipitous cliffs and sandy bays are nowhere surpassed for picturesque effectiveness.

3. Size. Shape. Boundaries.

Glamorganshire is situated at the south-east extremity of Wales, and is the most southerly of all the Welsh counties. It lies between 51° 24′ and 50° 48′ N. latitude, and between 3° 5′ and 4° 19′ W. longitude. Its boundaries are partly artificial and partly natural. On the south it is washed entirely by the waters of the Bristol Channel, which separates it from the opposite coasts of Somerset

Caswell Bay

and Devon ; on the north it is bordered by Breconshire and Carmarthenshire ; on the east the Rhymney river forms the natural line of demarcation between it and Monmouthshire; and on the west it is partly surrounded by the combined waters of the Bristol Channel and the Burry Inlet, and partly adjoins Carmarthenshire, from which it is divided by the Loughor river.

Except on the north its outlines are well defined, and its bold projection into the Bristol Channel gives it a marked individuality. The eastern boundary scarcely needs tracing in detail, for it follows strictly the course of the Rhymney river from its source, near Troed-y-Milwyr, the northern extremity of the county, to its mouth, two miles east of Cardiff. The western border-line may be described with equal brevity. It begins north-wards near Pantyffynnon in the valley of the Loughor, and descends the stream till it falls into the Burry Inlet at Loughor. The northern frontier lies amongst the mountains, and except for the fact that the border-line roughly corresponds to the northern limits of the South Wales coalfield, there are few natural features along its course to serve as landmarks, and it is chiefly an imaginary line across the hills ; but it occasionally acquires sharper definition by pursuing the beds of such highland water-courses as trend more or less eastwards and westwards. The chief valleys it presses into its service are the upper reaches of the lesser Taff, and the Cynon, the Sychnant Gorge, the Perddyn, and the upper Amman. It abruptly leaves the latter, however, at Cwm Amman, and cedes a rectangular corner to Breconshire by turning directly southwards to Nant Melyn, at the head of Cwm-y-Gors, and then pursuing its way across the mountains to the Cathan valley, which conducts it to meet the western boundary near Pantyffynnon. The places which roughly indicate its course from east to west are Troed-y-Melwyr, Pontsticill, Vaynor, Pant, Llwycoed, Hirwain, Pont Nedd Fychan, Gaer encamp-

Valley of the Rhymney
(*showing the Llanbradach viaduct*)

ment on the Perddyn, Ystalyfera, Brynamman, Cwm Amman, Nant Melyn, and Pantyffynnon. A line connecting these outposts (some of which lie just outside the county border) would roughly describe it.

In outline the county is exceedingly irregular, and in shape it somewhat resembles a shoulder of mutton with the knuckle pointing down channel, and representing the peninsula of Gower. Its extreme length from the Rhymney river near Ruperra to Worms Head is 54 miles, and its greatest width from Rhoose Point to Old Pitwell near Dowlais is 29 miles. It has a total area of 811 square miles or 518,865 acres; and in point of size ranks second amongst the Welsh counties, being exceeded only by Carmarthenshire. Since its limits were originally fixed by Act of Parliament in the reign of Henry VIII, its boundaries have remained unaltered.

4. Surface and General Features.

We generally speak of a county as flat, or mountainous, or diversified; but it is impossible to sum up the characteristics of Glamorganshire in any one comprehensive phrase. A glance at the map will show that it exhibits not only great irregularity of outline, but remarkable inequality of surface. Few counties show within similar limits such striking contrasts. So diverse are the features presented that different districts scarcely appear to belong to the same land. Soil, surface, flora, climate, and productions are all dissimilar, and even the inhabitants display divergent characteristics.

The county naturally falls into three well-marked divisions, whose differences are so striking that they have from time immemorial been regarded as three separate localities. There is a mountainous region in the north, an undulating champaign in the south, and an irregular and hilly peninsula in the west. The northern uplands were anciently called *Blaenau Morganwg*, the table-land in the south was named *Bro Morganwg*, and the western peninsula was spoken of as *Gwyr* or *Gower*. Bro Morganwg is still frequently referred to as *Y Fro* or the Vale. A line drawn from east to west across the centre of the county would divide the mountains from the plain.

Bishopston Valley

Gower might almost be regarded as a continuation of the plain out of which the waters of Swansea Bay have washed the connecting territory.

Bro Morganwg comprises a tract of undulating land some 10 miles broad by 22 miles long. It is bordered on the south by the Bristol Channel and extends northwards as far as Llantrissant; eastwards and westwards it stretches from the Rhymney to the Ogmore. Shut in between the mountains and the sea, it forms the Glamorganshire lowlands; but its designation of the Vale of Glamorgan is somewhat misleading, for it is in reality an undulating and hilly plateau ranging from 50 to 200 feet in altitude and rising in the centre to a table-land almost double that height. Seawards it terminates in a line of precipitous cliffs. It is watered by its own streams, which have scored its surface with a number of valleys, and it enjoys its own climate. Except for the commerce which collects round the seaports of Cardiff and Barry, it is almost entirely devoted to agricultural purposes, for which it is extremely suitable. It has been termed the "Garden of Wales." Though fairly well timbered, it is not sufficiently varied in feature to be strikingly picturesque, except in the centre, where the green monotony of the landscape is broken by the pleasant valley of the Daw. But what it lacks in scenery it makes up for by the number of its medieval antiquities. Nearly every village has the crumbling remains of a castle.

Blaenau Morganwg is very different, the transition from the plain to the highlands being exceedingly abrupt. The mountains rise like a wall round the northern edge of the Vale. Their general aspect is stern and forbidding, but the real beauty of the county was formerly to be found in their silent and solitary recesses. Everywhere the hills are furrowed with deep and secluded valleys which were once luxuriant with foliage and lively with the splash of falling torrents. Now everything is changed. The district has become one vast workshop, and the landscape is blurred with the smoke of innumerable collieries and furnaces.

The mountain system of Glamorganshire is extremely complicated. Broadly speaking all the Glamorganshire hills are buttresses to the still higher mountains of Breconshire. Though bold and precipitous they lack individuality, and baffle the observer by their multitudinous array. Their massed effect is, however, very striking. Roughly, these highlands fall into three principal groups, which for the sake of convenience may be termed the eastern, western, and central ranges. The central block is in plan a wedge-shaped mass of hills enclosed between the Nedd (Neath) and Cynon rivers. Right in the north at the apex of the triangle formed by these divergent streams rises the lofty summit of Craig-y-Llyn (1969 ft.), the monarch of Glamorganshire mountains, and radiating more or less southwards like the ribs of a fan from this common centre are a number of almost equally high spurs, Cefn Gwyngul (1489 ft.), Cefn Tadfernol (1692 ft.) and Cefn Rhondda (1567 ft.), Mynydd William Merrick (1769 ft.), Crug-yr-Afon and Mynydd Llangeinor (1859 ft. and 1755 ft.), Mynydd Caerau (1823 ft.), and Cefn Mawr (1560 ft.). On either side of this central cluster of mountains are a number of parallel ridges forming the eastern and western systems. Beyond the Cynon in the east, as we pass westward from the Rhymney, are Cefn Brithdir (1460 ft.), Cefn Gelligaer (1570 ft.), Cefn Merthyr (1292 ft.), and Mynydd Aberdare and Mynydd Merthyr (1346 ft. and 1336 ft.). On the west, on the other side of the Nedd, is the Graig-lwyd group, which at its highest point attains 1575 ft., Mynydd March Hywel (1371 ft.), Mynydd Alt-y-grug (1113 ft.), Cefn Gwrhyd (968 ft.), Mynydd-y-Garth (1057 ft.), Bryn Mawr (1153 ft.), Mynydd-y-Gwair (1226 ft.), and Graig Fawr (883 ft.). Though this is a tolerably complete enumeration of the various folds into which the northern

The Garth Mountain

surface of the county is crumpled, it is by no means an exhaustive catalogue of all the hills which the county contains. Two eminences which are very conspicuous from the lowlands should be mentioned. They are the Garth mountain (1009 ft.), which lies to the north-west of Cardiff, and Mynydd Margam (1409 ft.), a very interesting hill which rises immediately behind the village of that name.   The summits of all these hills are wild and desolate, and their bleak solitudes offer a striking contrast to the teeming valleys at their feet.   Though a source of wealth to the mine owner, they are used by the farmer chiefly as sheep-runs.   They abound everywhere in prehistoric antiquities.

Gower, our third division, forms the western portion of Glamorganshire, and in its general features is sharply

Mewslade Bay

differentiated from the rest of the county.   It exhibits, however, the same characteristic contrast between hill and plain.   There is a mountainous district in the north which falls within the limits of the coalfield, and is now one of the most populous and busy localities in the shire ; and there is a sea-bound pastoral district in the south.   The name is now generally restricted to this rocky peninsula which forms such an eccentric termination to the county

on the west. It is a wild and solitary land almost surrounded by the sea. In measurement it is 18 miles long and
8 miles broad, and has an acreage of 80 square miles. It is
chiefly remarkable for its bold and diversified coast scenery.
Few districts can show such an extensively serrated shore,
for it has 45 miles of sea-board. Compared with the
varied beauty of the cliffs, the interior is somewhat bald
and featureless. A low ridge of hills—Cefn-y-Bryn—
some 600 feet in height, runs diagonally across the peninsula, and in the extreme west rises a ridge of bare and
breezy downs of equal altitude. Though woods occasionally clothe the more sheltered sides of the cliffs,
Gower as a whole is treeless. With the exception of
one or two pretty glens its valleys and watercourses are
few. Agriculture is the prevailing occupation, but the
fields yield scanty crops compared with the prodigal
harvests in the Vale of Glamorgan. It is in consequence
a sparsely populated region, and the villages are remote
and little visited except by tourists. Like the rest of the
county the peninsula abounds in antiquities.

5. Geology and Soil.

The geological character of Glamorgan is comparatively
simple. All the rocks have been formed by the action
of water ; there are none of igneous origin. Apart from
the alluvial deposits along the estuaries of the rivers, there
are only five systems represented within the county, the
Silurian, Devonian, Carboniferous, Triassic, and Jurassic.

	NAMES OF SYSTEMS	SUBDIVISIONS	CHARACTERS OF ROCKS
TERTIARY	Recent Pleistocene	Metal Age Deposits Neolithic „ Palaeolithic „ Glacial „	Superficial Deposits
	Pliocene	Cromer Series Weybourne Crag Chillesford and Norwich Crags Red and Walton Crags Coralline Crag	Sands chiefly
	Miocene	Absent from Britain	
	Eocene	Fluviomarine Beds of Hampshire Bagshot Beds London Clay Oldhaven Beds, Woolwich and Reading Thanet Sands [Groups	Clays and Sands chiefly
SECONDARY	Cretaceous	Chalk Upper Greensand and Gault Lower Greensand Weald Clay Hastings Sands	Chalk at top Sandstones, Mud and Clays below
	Jurassic	Purbeck Beds Portland Beds Kimmeridge Clay Corallian Beds Oxford Clay and Kellaways Rock Cornbrash Forest Marble Great Oolite with Stonesfield Slate Inferior Oolite Lias—Upper, Middle, and Lower	Shales, Sandstones and Oolitic Limestones
	Triassic	Rhaetic Keuper Marls Keuper Sandstone Upper Bunter Sandstone Bunter Pebble Beds Lower Bunter Sandstone	Red Sandstones and Marls, Gypsum and Salt
PRIMARY	Permian	Magnesian Limestone and Sandstone Marl Slate Lower Permian Sandstone	Red Sandstones and Magnesian Limestone
	Carboniferous	Coal Measures Millstone Grit Mountain Limestone Basal Carboniferous Rocks	Sandstones, Shales and Coals at top Sandstones in middle Limestone and Shales below
	Devonian	Upper } Mid } Devonian and Old Red Sand- Lower } stone	Red Sandstones, Shales, Slates and Lime- stones
	Silurian	Ludlow Beds Wenlock Beds Llandovery Beds	Sandstones, Shales and Thin Limestones
	Ordovician	Caradoc Beds Llandeilo Beds Arenig Beds	Shales, Slates, Sandstones and Thin Limestones
	Cambrian	Tremadoc Slates Lingula Flags Menevian Beds Harlech Grits and Llanberis Slates	Slates and Sandstones
	Pre-Cambrian	No definite classification yet made	Sandstones, Slates and Volcanic Rocks

But the distribution of these various formations is so extremely unequal that the list may be virtually reduced to two—the Carboniferous and the Jurassic. Of these two the Carboniferous so largely predominates that seven-eighths of the surface of the shire comes under this classification. It will be seen that the geological map in its main features roughly follows the geographical divisions of the county. The whole of the northern highlands consists of Coal Measures; the Vale of Glamorgan is Lias; and the peninsula of Gower is Mountain Limestone. The Silurian, Devonian, and Triassic formations are only sparsely represented. Their rocks occur in patches, and form no very striking feature in the geological field. Taken in the order of their formation, the various systems will be found distributed over the following areas :—

The Silurian rocks, though occurring in such large masses elsewhere in South Wales, are only very scantily exhibited in Glamorganshire. They are found on the extreme eastern verge of the county and nowhere else. On the banks of the Rhymney there is a small bed of shales, sandstones, and mudstones, which projects across the border from Monmouthshire.

The Old Red Sandstone, which forms the bulk of the adjoining counties of Breconshire and Monmouth-shire, continues as far as the river Taff, dips right under Glamorganshire in a south-westerly direction, and only reappears within the county as a small outcrop between Cardiff and Bridgend, and as a range of hills running across the surface of the peninsula of Gower. The

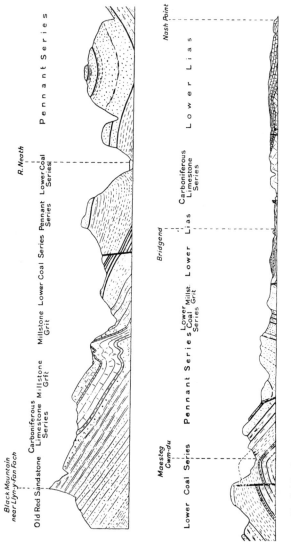

Geological section from the Black Mountain in Carmarthenshire to the Sea near Bridgend

depression caused by the disappearance of the Old Red Sandstone forms a trough in which the Coal Measures lie.

The Carboniferous system is the prevailing formation of the county. All the massive hills of the north belong to it, and the vast area over which it extends is the cause

Sandstone at Fairoak Farm, Roath Park

of the great prosperity of the shire. The coal-bearing rocks, though lying above the Old Red Sandstone, do not rest immediately upon it. They are underlaid everywhere by an intermediate bed of Mountain Limestone and Millstone Grit, which are also classified as Carboniferous

rocks. Prior to the Carboniferous epoch, the Red Sandstone hills of Breconshire formed the shore of some shallow
and clear sea, and Glamorganshire was a submerged ledge
of rock upon which a multitude of coral animals subsequently laid down this thick floor of limestone. The
limestone bed not only underlies the whole of the
Glamorgan coalfield, but forms a thick rim round its
southern edge. The deposit varies from 500 to 1000
feet in thickness. The rock is found elsewhere, besides
at the bottom of the coal basin. It occurs immediately
below the Lias in some portions of the Vale, where it
occasionally breaks out on the surface. It forms some
of the most prominent points along the coast, and is
exposed as a long stretch of low rocks between Porthcawl
and Sker Point. It again appears as the predominant
geological feature in the peninsula of Gower, where it
can be seen upheaved at a considerable angle in the bold
cliffs which form the coast. Some of the cliffs in this
district are characteristically perforated with caverns,
which on their first exploration were found thickly strewn
with the bones of extinct animals.

Between the Mountain Limestone and the Coal
Measures is a deposit of Millstone Grit varying from
250 ft. to 1000 ft. in thickness. It comes to the surface
at the foot of the hills stretching as a thin tract of
territory between Caerphilly and Llantrissant. It appears
again as a longish patch north of Bridgend. It is a coarse
quartzose sandstone frequently used for millstones, but
locally known in the coalfield as the "Farewell" rock.
Its general position in Wales is at the base of the Coal

Measures, and the miner who reaches it regards it as a signal to abandon further search.

The South Wales coalfield stretches from Pontypool to St Bride's Bay, and has an area of 1000 square miles, of which Glamorganshire possesses about half. The extreme length is 90 miles, and the breadth varies from 21 miles in Glamorganshire to 1½ miles in Pembrokeshire. It rests in a sort of pear-shaped trough with the widest end towards the east. The greatest depth at which coal lies is calculated to be about 6650 feet. Across the basin run two anticlinal folds, one stretching from Risca to Aberavon, and the other from Cwm Neath to Kidwelly. Swansea Bay makes a large encroachment upon the Glamorganshire coalfield, and a colliery near the shore at Port Talbot is worked beneath the Channel. Though the Coal Measures have an estimated thickness of 7000 ft., the actual seams of coal vary only from 1 to 6 ft. thick, and are thinnest and most numerous in the west. The seams are divided one from another by intervening masses of sandstone, which in some cases are as much as 500 ft. thick. The layers of sandstone are again interspersed with deposits of shale, which measure in thickness from 10 to 50 feet. Most seams of coal rest upon a co-extensive cushion of clay, which varies from 6 inches to 10 feet in depth. This underclay is in South Wales, though not elsewhere, an invariable accompaniment of the coal, and is believed by some to be the soil in which the vegetation forming the coal originally grew, as in some cases it contains rootlets. It is a soft sandy shale extensively used in the manufacture of fire-bricks.

The absence of soda and potash salts, possibly abstracted by the plant life which once flourished above it, preserves it from fusion when exposed to the action of heat.

The coal seams, which lie in three series, are known as the Upper Coal Measures or Upper Sandstones, the Pennant Grit, covering most of the coalfield, and the Lower Coal Measures. In the Upper Coal Measures the coal is highly bituminous and inflammable, and is used chiefly for domestic purposes, and for the production of coal gas and coke. In the Pennant Grit series the coal is much less plentiful and is sandwiched in between beds of hard sandstone, which are frequently quarried for building and paving. This coal readily cakes and is employed largely in the manufacture of "patent fuel." The Lower Coal Measures produce the famous steam coal, so invaluable for boiler furnaces. It contains a much larger percentage of carbon than the inflammable varieties, and produces in consequence a much fiercer heat and less smoke. Glamorganshire supplies virtually all the larger navies of the world with their fuel. As the coal seams spread towards the north-west the coal becomes still less bituminous in character, and to the north of Swansea in the valleys of the Tawe and Twrch it is almost purely anthracitic and smokeless.

Though some violent upheaval has lifted the Coal Measures to their present position as lofty hills, there is little doubt that they were first laid down at the bottom of some lagoon into which a number of muddy streams were constantly discharging a large quantity of earthy sediment. Rank vegetation sprang up on the more elevated portions of these swamps, which, owing to their unstable nature, subsequently sank. The plants which covered these spongy deposits consisted chiefly of gigantic mosses and ferns. The forests so formed disappeared or were renewed as the land fell or rose. The mud eventually hardened into shales, and the coal seams are the compressed remnants of these vanished jungles. Some violent shrinkage of the earth's crust in ages long subsequent to its formation raised the coalfield to its present altitude. The valleys which score the surface of the hills are not the folds into which the hills were thrown, but the furrows ploughed out of their surface by the torrents which coursed down their sides. The streams which now wind their way at the bottom of the gorges are the shrunken survivals of these ancient rivers. The Lower Coal Measures are rich in ironstone and contain a number of marine shells and Entomostraca. In the Upper Measures the only fossil found, other than the remains of plant life, is the shell *Anthracosia*.

The Triassic rocks which occur within the county are found chiefly as patches of dolomitic conglomerate near Llandaff, St Fagan's, Coychurch, Pyle, and Newton Nottage, and are overlaid by Keuper Red Marls in the neighbourhood of Bridgend. In the conglomerate of Newton Nottage have been found the footprints of some gigantic three-toed bird or reptile. A slab from the district bearing this impression is now exhibited in the Cardiff Museum. The geological formation of Penarth Head, where the rocks also belong to this series, is extremely interesting. At the base of the cliff is a foundation of red Triassic marls, and upon these reposes a 28-foot bed of green Rhaetic marls. Above this may be seen a larger bed of black shales, 24 feet thick, containing a bone-bed rich in the scales and teeth of fishes and reptiles. This is capped, as the coast trends towards Lavernock Point, with a deposit of white Lias, 18 feet in

Penarth Cliffs

thickness, and composed of grey and brown sandy shales. The marls are quarried in the neighbourhood of Llandough and are ground up for brick-making.

The Lias of Glamorganshire and Monmouthshire is represented by the Lower Lias only, and borders the southern part of these counties and the South Wales coalfield. It occurs more or less in patches, the greatest development being in the district known as the Vale of Glamorgan, between Barry and Bridgend, where remarkably fine sections can be seen in the cliffs along the seashore.

All along the coast from Lavernock Point to the mouth of the Ogmore river the Lower Lias shales and grey limestone rest upon the Rhaetic beds, and constitute the long series of cliffs which form the shore line of the Vale of Glamorgan. These rocks extend inwards as far as Cowbridge, and compose the floor of the Vale. At Southerndown they form a conglomerate which represents an ancient beach, and may be seen resting on the upturned edge of the Mountain Limestone. As may be observed from an inspection of the cliffs the stratification of the Lias rocks is extremely regular. The rocks consist of bands of blue limestone six inches in thickness, between which are interposed layers of shale, sometimes called, on account of their perfect lamination, paper shales. The White Lias series contain a large number of fossils, including shells and the star-fish *Ophiolepsis Damesii*. From the Blue Lias rocks have been obtained more than fifty species of corals, and a great many Ammonites. Saurian remains are also not infrequent, Ichthyosaurus and Plesiosaurus being found in the large quarries attached to the cement works at Penarth.

As one travels westward the thickness of the Lias increases. The total depth in the district round Newport is only about 30 feet, at Penarth and Lavernock the deposits attain a thickness of about 130 feet, but in the neighbourhood of Aberthaw they reach several hundred feet.

Section of the Lower Lias at Lavernock

Generally speaking, the deposits consist of layers of blue or blue-grey argillaceous limestone alternating with blue, white, or yellowish shales and clays, the blue shales often containing notable quantities of iron pyrites. In the east the clays and shales predominate and the deposits attain their most aluminous character; westward they become more siliceous and also more calcareous, the beds of stone greatly predominating.

Surface deposits of alluvial matter occur at the estuaries of all the rivers and along the valleys of the Nedd and Tawe. Extensive tracts of marshy land have been formed between the mouth of the Rhymney and Penarth Head, and at the foot of the hills in the neighbourhood of Aberavon. There is a patch of undrained bog at Crymlyn between the estuaries of the Nedd and Tawe, and some extensive salt marshes line the southern shores of the Burry Inlet. The movements of glaciers in the county have left some traces. The accumulations of sand and gravel which are found in most of the river valleys, though generally attributed to excessive rainfalls in prehistoric times, may really be of glacial origin. There is a glacial drift near Cardiff consisting of pebbles and boulders of Old Red Sandstone and other rocks lying immediately to the north. A number of "erratics" from the local Carboniferous Limestone and Millstone Grit are scattered about the Ely valley. But the most interesting deposit is at Pencoed, where well-polished boulders of rocks foreign to the district have been found, which microscopic examination has shown to have been brought from North Wales by glacial action.

Large boulders of Millstone Grit or Conglomerate are also frequently met with in Gower.

Iron ore is found in the form of haematite in deposits amongst the Mountain Limestone, and a great deal of ironstone is obtainable amongst the lower Coal Measures. The chief iron-ore district lies in the neighbourhood of Merthyr, Dowlais, and Aberdare; and deposits have been worked near Pentyrch. Gypsum occurs in small beds near Penarth.

The soil of the different localities varies very considerably both in character and value. In the Vale of Glamorgan it is a deep rich loam of unusual fertility and light to handle. It is peculiarly suitable for the production of wheat. The underlying substratum of limestone adds considerably to its quality, but gives it a very stony appearance when turned over. The north, on the other hand, is a very barren region. The damp peaty earth with which the mountains are thinly covered is very poor, and the drier patches are too gravelly to be of any service to the agriculturist. In Gower again the soil, though it varies a good deal, as a rule lacks both depth and virtue. It ranges from a loose red earth to a thick yellowish clay, and is occasionally interspersed with beds of sand. In general it yields only a moderate return for the labour involved in its tillage.

6. Watershed and Rivers.

Glamorganshire is particularly well watered. It abounds in rivers, though none of them are of any great size. They are all short, and take their rise either within the county or else amongst the Brecon hills just beyond its borders. The watershed lies entirely in the north, and all the streams run southwards and empty themselves into the Bristol Channel. They were at one time distinguished for the picturesqueness of their surroundings, but their beauty has now disappeared before the pitiless ravages of industrial progress. None of the rivers are in themselves of much value as waterways, for they are narrow, rapid, and shallow. They have nevertheless played an important part in the industrial development of the county, for their valleys are the commercial arteries of the district, and the estuaries into which they empty themselves have provided the harbours which now place the mineral wealth of Glamorgan at the disposal of the world. The rivers may be conveniently regarded as falling into three distinct groups, which form the eastern, western, and central drainage systems of the county. The easterly basin comprises the Rhymney, Taff, and Ely rivers, which flow in a south-easterly direction. The westerly basin consists of the Nedd, Tawe, and Loughor, which flow south-west. The central basin is drained by the Ogmore, Kenfig, and Avon, the first two of which flow southwards, and the last south-west. A subordinate group of streams, of which the principal constituent is the Daw, drain the Vale of Glamorgan.

The Rhymney has its birth in Breconshire, but for the remainder of its career it forms the dividing line between Monmouthshire and Glamorganshire. Receiving the Bargoed brook on its way, it follows a fairly consistent course southwards; but, before it eventually reaches the Channel, it has to make an awkward détour to the east to work round the obstructing ridge of Cefn Carnau. The valley through which it flows, though exceptionally rich in minerals, is bare and otherwise lacking in interest. At Caerphilly, however, there stands at the entrance of an adjoining vale one of the most famous of the ruined castles of Wales. Below Caerphilly the scenery improves until, leaving its gorge, the river twists itself as a muddy estuary across the flats of Cardiff.

Unlike the solitary Rhymney, the Taff is a veritable family of rivers. Rising as a twin stream in Breconshire, it follows a double course as far as the large industrial town of Merthyr, where the greater Taff and lesser Taff unite. At Quaker's Yard the river receives a further accession to its waters in another Bargoed brook, and at Abercynon its volume is again swollen by the Cynon river, which likewise descending from the Brecon highlands flows past the large and populous towns of Aberdare (which stands on one of its feeders) and Mountain Ash. Another stream falling into the Cynon before it joins the Taff is the Aman; and the Clydach, which waters a very busy little glen, reaches the Taff later. At Pontypridd the Taff is joined by two other important tributaries, the Rhondda Fawr and the Rhondda Fach. Both of these streams have their source in the spurs of Craig-y-Llyn, and pass through one of the most populous districts in South Wales. The valleys through which they flow contain some of the finest steam coal in the world, and they have in consequence of its discovery undergone a remarkable transformation. The two Rhondda rivers coalesce at Porth, and empty their united current into

The Taff

(showing the Garth in the distance)

the Taff at Pontypridd. The position of Pontypridd at the confluence of these valleys makes it a sort of general clearing-house for the vast traffic of the district. It borrows its name from a remarkable one-arched bridge which a local mason named Edwards built in 1750 across the Taff. Five miles above Cardiff the Taff emerges from the mountains and makes its way across the level plain to the sea. Before the river reaches Cardiff it glides past the peaceful city of Llandaff, whose rural seclusion conceals its long and chequered history. The cathedral is pleasantly situated on the banks of the stream. The wide tidal estuary by which the Taff finally discharges its waters into the Channel has given Cardiff the opportunity of becoming one of the largest ports in the kingdom.

The Ely river is a short stream rising amongst the hills behind the mining village of Ton-yr-Efail. For a short distance it wanders amongst the mountains, and then uniting itself with the Mychedd brook it creeps through a narrow pass near Llantrissant (Llantrisaint) into the Vale of Glamorgan. The town of Llantrissant is perched on a knoll above the river, and its position at the mouth of the defile made it in medieval times a place of much strategical importance, and its castle was strongly fortified. Below Llantrissant the Ely receives the Afon Clun, and then meanders southwards across the Vale to Peterston, where there are the remains of another castle. At Peterston it suddenly turns eastwards, and flowing past the castle and battle-field of St Fagan's reaches Ely, where it again bends southwards and finally finds its way into the estuary of the Taff at Penarth, where it forms a natural harbour. Some fine docks have been constructed beneath the headland to supplement the commercial conveniences of the river.

The rivers of Mid-Glamorgan, like their eastern and western neighbours, run a short impetuous course amongst the hills before they seek a more sluggish channel in the plains on their way to the sea. In number they are three—the Ogmore, the Kenfig, and the Avon. The Ogmore collects together a number of converging streams which have their origin amongst the cluster of lofty hills which are massed together in the centre of the county. The Ogmore proper, like the Taff, is a combination of dual streams, the Ogmore Fawr and the Ogmore Fach, which rise on either side of Carn Fawr (a spur of the central hills) and unite at Blackmill. At Abergarw the joint river receives the turbulent waters of the Garw, which comes rushing down the valley between Mynydd Llangeinor and Mynydd Caerau. At St Bride's it effects a junction with a less impetuous but more important tributary, the Llynfi, which rises on the western side of Mynydd Caerau, and passes the busy mining centres of Maesteg and Tondu. Thenceforth the Ogmore flows a placid stream through the town of Bridgend, below which it unites with the waters of the Ewenny, and then empties itself into the Bristol Channel near Sutton by an estuary remarkable for its vast accumulations of sand. The old priory church on the banks of the Ewenny, and the crumbling ruins of Ogmore Castle near the estuary, invest it with considerable antiquarian interest. The Alun brook, which is a tributary of the Ewenny, cuts its way through a small but singularly picturesque gorge to join the larger stream, and furnishes the landscape with a very striking and unexpected feature. The Kenfig is little better than a streamlet, and descends from the steep slopes of Mynydd Margam to push its way to the sea through the sandy wilderness known as the Kenfig Burrows. The Avon is a larger river, which rises as a mountain brook on the slopes of Crug-yr-Afon, and joining itself to the Corwg—a stream descending from the lofty spurs of Craig-y-Llyn—strikes out a south-westerly course for Swansea Bay. Before discharging its waters here at Aberavon it collects two other tributaries, the Pelena and the Dyffryn. The course of the river is walled in all the way by bare and rugged

Ogmore Castle

hill sides, and its banks are lined with collieries and iron-works, which send down their products for shipment to the extensive dock at Port Talbot at the mouth of the river.

The rivers of West Glamorgan are the Nedd, the Tawe, and the Loughor, and they all flow more or less in a south-westerly direction. The Nedd, the most famous of the three, may be said to be the queen of Glamorganshire rivers. It wins its popular reputation, however, chiefly before it enters the county, when still a mere cluster of mountain burns. The four streams which unite to form the "full fed" river—the Hepste, the Mellte, the Nedd, and the Perddyn—all rise amongst the solitudes of the Breconshire Vans, and are celebrated not only for their wild and romantic surroundings, but for the numerous cascades in which they abound. But if these earlier beauties belong to Breconshire, Glamorganshire can still claim some share in making the river what it is, for it furnishes for its channel a still beautiful, but sadly tarnished valley. The course of the river is short. Collecting its scattered sources together it enters the county at Pont Neath Vaughan and flows with a fairly straight course south-west into Swansea Bay. Passing on its way the mining villages of Glyn Neath and Resolven it allies itself at Aberdulais with its most considerable tributary, the Dulais, which flows down to join it from the heights of Mynydd-y-Drum, and below the town of Neath it is further augmented by the Clydach. On its right bank a little below the town stand the grimy ruins of Neath Abbey. At the busy little port of Briton Ferry the valley again narrows and the river makes its escape into Swansea Bay through a bold and picturesque defile. Two small waterfalls occur on the course of the Nedd, and on the Melincourt brook near Resolven there is a slender cataract some 80 feet in height.

The Tawe also rises amongst the Breconshire Vans, and pursuing a course almost parallel with that of its sister river the Nedd, falls into Swansea Bay at Swansea. Its chief tributaries are the Twrch—an impetuous stream which comes down from the Black Mountains and meets it at Ystalyfera—and another Clydach which joins it at Clydach. The Twrch marks practically the northern limits of the coalfield, and in its valley are situated most of the valuable anthracite seams. The valley of the Tawe is now a scene of industrial desolation crammed with smelting works. The capital of this sulphurous region is Landore. At Swansea the estuary of the Tawe forms a very fine entrance to the large and spacious docks which belong to the town.

The Loughor is the only Glamorganshire river which rises in Carmarthenshire, and it remains in part a Carmarthenshire stream throughout, for it forms the boundary between that county and Glamorganshire. The only contribution which Glamorganshire makes to its waters is the little river Dulais, which rises on the slopes of Mynydd-y-Gwair, but another stream, compounded of the Afon Lliw and the Afon Llan, falls into its estuary below the town of Loughor. Some miles before the Loughor joins the sea it forms a long and shallow lagoon which abruptly narrows at Loughor town, and opening again into a wide and sandy estuary changes its name to the Burry Inlet.

The streams which water the Vale of Glamorgan are the Daw river and the Cadoxton and Colhugh brooks. The Daw runs a short but very picturesque course from Llansannor to the sea. Passing the old-fashioned town of Cowbridge it winds its way through a wide and prettily wooded vale, which is sprinkled with the remains of ancient castles. Picking up the Llancarfan brook, which scoops its way through a deep defile past the site of the ancient monastery of Llancarfan, it finally empties itself into the sea by a wide and marshy estuary at Aberthaw. Aberthaw was once a small port, but is now only a village. An ancient harbour is said to have once existed also at the mouth of the Colhugh brook, which flows into the Channel below Llantwit Major, but all trace of it has vanished. The business which has forsaken the Daw and the Colhugh has, however, come to the Cadoxton river, for at its mouth have been created the gigantic docks of Barry.

The streams which drain the peninsula of Gower are few and trivial. They are the Clyne river, which flows across the eastern end of the peninsula and falls into Swansea Bay; the Ilston brook, which waters the centre and empties itself into Oxwich Bay; and the Burry brook in the western extremity, which flows northwards and joins the Burry Inlet near Cheriton.

7. Natural History.

Geology shows us that the surface of the earth has at various times undergone alterations and changes of a stupendous nature. Great mountain chains have been elevated, new seas have been formed, vast land-masses have disappeared. Some of these changes are exemplified in our own country. We know, for example, that at one time—and that, geologically speaking, at no very remote period—Britain was not insular, but formed part of the European continent.   In the caves of Derbyshire, Devon, Gower, and many other places are found the bones of extinct animals identical with those disinterred in neighbouring continental countries, or dredged up from the bed of the North Sea.   Again, in many places around our coasts are to be seen, at extreme low water, the remains of forests buried beneath the sea.   Though just off the west coast of Ireland the sea bottom sinks rapidly to very deep soundings, the North Sea is everywhere very shallow, and if London could be placed in it, the dome of St Paul's would be seen standing well above its surface.   Great Britain and Ireland are thus examples of what are known to geologists as "recent continental islands," and subsidence, erosion, and other geological changes have turned dry land into the North Sea, and broad river valleys into the Bristol and English Channels.

But, at some earlier period, when still forming part of the continent, our land was for a time submerged.   The existing fauna and flora destroyed by this subsidence had thus to be replaced from the continental lands lying to the south-east.   Slowly these new immigrants worked their way north-westward as the land again rose and afforded suitable conditions, but as it was not long before separation occurred and Britain became insular, not all the species existing on the continent were able to establish themselves in our land.   We should thus expect to find that those parts of the country nearest the continent were richer in species and those furthest off poorer, and this proves to be

the case both with plants and animals. Britain has fewer
species than France and Belgium ; and Ireland, which
was probably separated still earlier, has fewer than
Britain.

Both plants and animals depend for their existence
upon suitable climatic conditions and upon a favourable
environment. The British Isles are not of sufficient
extent to exhibit great climatic diversity, but they have
considerable variety of physical feature. The consequence
is that much the same plants and animals are to be found
in different parts of the kingdom where the same physical
conditions prevail. The absence of any particular species
in a locality where it might naturally be looked for, is
generally due to accidental or artificial causes. Other
things being equal, those places enjoy the greatest advantage from a naturalist's point of view which lie nearest to
those regions whence the original stock was derived. As
the stream of migration was from the south-east, the
southern counties are usually the richest in objects of
natural history. Glamorgan occupies a fairly good
geographical position in this respect, and its physical
peculiarities have made it a suitable habitation for the
many different kinds of wild life which have from time to
time reached its shores. Mountain, plain, and seashore
all have their own special denizens, and there are still
large tracts of uncultivated and sparsely-populated territory,
where they can flourish undisturbed. In one respect the
naturalist has reason to complain that the county has been
unfortunate. Some of its most secluded recesses, once the
favourite haunts of bird, beast, and flower, have in recent

years undergone vigorous commercial exploitation, with the result that many interesting things have been destroyed or greatly reduced in numbers. The streams especially have suffered much by the pollution of their waters, and the fish for which they were once famous have disappeared. Generally speaking, however, the fauna of Glamorganshire is well up to the average in interest. Most of the species met with in other counties are represented somewhere within its borders. The red deer, however, once common and still found in the opposite highlands of Somerset and Devon, does not now roam the hills of Glamorgan. The wild cat has been exterminated, the polecat is rare, the otter has been greatly reduced in numbers, and the badger now is not very often met with. The seals which once frequented the shores of the Severn for the sake of the salmon have abandoned their visits. Seven, at least, of the twelve species of English bat are found in the county. There are the usual common rodents, but the hare is scarce, and the water vole is not common. The brown rat, unknown in England before the beginning of the eighteenth century, is of course the common species of the county, but the black rat still occurs.

Birds, except in the colliery districts, where they find little shelter and many enemies, are fairly numerous, though they only exist in any very great variety in special localities. The singularly diversified character of the county is favourable to a miscellaneous bird life, but the destruction of much of the woodland has deprived it of some of its earlier wealth in this respect. Of the birds of prey the peregrine falcon, the buzzard, and the long-eared owl are occasional visitants. The merlin is sometimes seen. The greater and lesser spotted woodpecker breed in the county, but not in great numbers. The nightingale is reported to have been sometimes met with, but all attempts to naturalise it have failed. The sea coast is the most interesting district to the ornithologist. The

Kenfig Pool

neighbourhood of Worms Head in particular is the breeding ground of a prodigious number of sea-fowl, which nest in these solitudes. Amongst the less common kinds of waders and swimmers to be met with are the great crested grebe, the great black-backed gull, and the stormy petrel. The cormorant is frequently to be observed perched on the ledges of the cliffs, but the bittern, once fairly common, is now very rare, and only comes during hard winters.

The rivers have been almost entirely denuded of their former occupants. Even dace and chub have disappeared. Curious changes in the live-stock of streams and pools sometimes occur from natural causes. Kenfig Pool was once noted for the immense quantity of pike it contained, but none are to be found there now. The naturalist will however find in the products of the shores some compensation for what he has lost in the rivers. The sandy creeks and rock pools of the Gower peninsula in particular will provide him with numerous and varied specimens of marine life. The curious little blenny, which spends half its time out of the water, is occasionally found in Swansea Bay, and there, too, are sometimes captured the sword-fish, the globe-fish, and the flying-fish. The limestone coast of Gower abounds in crustacea and mollusca of all kinds.

The entomologist will find plenty to interest him in Glamorganshire. The warrens of Merthyr Mawr, the Crymlyn burrows and bog, and the sandhills of Oxwich Bay have always been the favourite preserves of the insect hunter. Butterflies and moths in great variety are to be found in the Vale of Glamorgan and in the Gower peninsula. Some of the rarer specimens, however, only make their appearance during favourable summers.

The flora of Glamorgan is much more abundant and varied than the fauna. Some species common here are only of local occurrence elsewhere. The plant life of a locality is determined by its soil as much as by climatic and physical considerations. Both the diversity of its physical structure and its mixed geological character have made the county peculiarly rich in the variety of its botanical specimens. In descending from the hills to the plains, or to the rich and sheltered valleys, one is struck at once by the luxuriance of the vegetation displayed. It is like coming from a desert to a garden. The peculiarities of the maritime region are again very discernible. In approaching the coast one becomes aware of the proximity of the sea long before it is reached. Trees are scarce and stunted. A new set of flowering plants replaces the old favourites of wood and meadow. Grasses, reeds, and rushes straggle along the sand and pebbles. The influences of soil are equally marked. The elevated limestone tracts around Morlais Castle show a much more exuberant plant life than the ridges of Pennant Grit on the other side of the valley of the Taff; and the flora of the hard limestone headlands of Gower is distinct in many particulars from that of the softer lias rocks which fringe the Vale of Glamorgan. Even within the same geological area differences can be noted. The white down of the cotton grass shows in great patches on the boggy sides of the hills but does not appear on their rocky summits ; and the grasses and lichens which cling to the cliffs within reach of the spray are very different from the dwarf shrubs and rushes which cover the sand dunes, or from the plants on the salt marshes overflowed by the tide. Roughly speaking there are three well-defined botanical zones—the hills and valleys of the Coal Measures, the Vale, and the sea coast, each with its own characteristic vegetation. The mountains frequently exhibit a sub-alpine element in the flora, and the valleys are specially distinguished by the richness of their bramble growths. These are extraordinarily abundant : of the 190 forms of Rubus known to Britain, nearly 60 are found in the Rhondda and Cynon valleys. The Vale is the garden of Glamorgan. In the spring the banks and meadows are a blaze of colour. Here lime-loving plants abound, and *Clematis vitalba* is a great feature of the hedges and woods. The forest trees chiefly seen in the fields are the ash, beech, elm, and sycamore, whilst the oak, birch, alder, fir, and larch clothe the sides of the hills. The wych elm is believed to be a native in Glamorgan and is said to stand the sea air better than any other tree except the sycamore, which likewise flourishes freely.

The plants of the sea-coast, though not so showy, are, however, botanically the most interesting. The marshes, bogs, sand and shingle, and the lias and limestone cliffs, present a flora of great richness and variety. The wind-tossed sandhills near Kenfig and Merthyr Mawr are carpeted with miniature briars and brambles, and nowhere does the viper's bugloss grow as it does on these dunes. About Nash Point is to be seen the tuberous thistle, *Cnicus tuberosus*, elsewhere native only in Wiltshire, and near the coast the rare fen-orchid (*Liparis Loeselii*) has been met with. The peninsula of Gower may be almost regarded as a region apart, so special is its vegetation. It is a botanist's paradise. Here is found the one species in these islands peculiar to Glamorgan, *Draba aizoides*, the yellow "whitlow-grass" of the Alps and Pyrenees.

Its cousin, *Draba* (*Erophila*) *verna*, is more common. At Park Mill is the hairy cress, *Arabis hirsuta*, and on the sands in the neighbourhood appears the sea heron's-bill, *Erodium maritimum*, and occasionally the sea-barley or squirrel-tail, *Hordeum maritimum*. The yellow rock-rose, *Helianthemum canum*, is found on Cefn-y-Bryn and Worms Head, and the meadow clary (*Salvia pratensis*) and *Osmunda regalis* are amongst the less common plants to be met with in the peninsula. Sometimes a foreigner has been introduced into the county by accident. One of the oraches, *Atriplex pedunculata*, has been discovered at Aberavon, borne thither probably in ballast.

The shores of Glamorgan are not as a rule so richly draped with seaweed as other parts of the coast. *Ulva porphyra*, however, is plentiful off Gower, and is turned by the natives into laver bread, and together with samphire (*Crithmum maritimum*) is sold as an edible commodity in Swansea market.

8. A Peregrination of the Coast.

The seaboard of Glamorganshire is remarkable both for its extent and variety. Throughout the 88 miles of frontage which it exposes to the sea it can furnish an illustration of almost every possible coastal characteristic. But though its features are so varied they lie for the most part grouped together in fairly uniform sections. Roughly speaking there are three distinct districts—the rock-bound fringe of the Vale of Glamorgan, the sandy shores of

Swansea Bay, and the rugged and serrated peninsula of Gower. The scenery, though nowhere wanting in interest, is artistically unequal. It improves as one proceeds down channel. The long lines of cliffs which for the most part border the plain are almost monotonous in their regularity, and lack both the background of encircling mountains which give such charm to the fine panorama of Swansea Bay, and the romantic fascination of the wild and jagged rocks of Gower. It is impossible to do more than give a very sketchy outline of this extensive and interesting region.

From the mouth of the Rhymney, whose tortuous estuary separates the coast of Glamorganshire from that of Monmouthshire, a long and featureless expanse of alluvial deposit stretches to the mouth of the Taff, round which is gathered the great city and port of Cardiff. Extensive docks have been cut in the moors, and a forest of factory chimneys mingles in the picture with the masts of the shipping in the harbour. On the other side of the estuary of the Taff, which also receives the waters of the Ely river, rises the fine headland of Penarth. It stretches like a gigantic wind-screen across the mouth of the port, and beneath its shelter vessels can ride at anchor in the roads in perfect security. Some smaller docks at Penarth lie immediately below the headland and furnish further accommodation for the immense business of the port. The headland carries no lighthouse, but the church on the summit of the hill is a well-recognised landmark both with the sailor and the landsman. Penarth, besides being a favourite residential suburb of Cardiff, has become something of a watering-place and possesses a pier and promenade.   Prominent though Penarth Head is, it is not seen by vessels coming up channel, for the coast trends for 2½ miles directly southwards to Lavernock Point before it bends westwards.   The cliffs between these two promontories are of great geological interest.   Lavernock Point, like Penarth Head, is unlighted, for the channel is

Penarth

sufficiently illuminated by a lighthouse on the Flat Holm, which stands in mid-channel three miles south-east of the point, and forms an appendage to the port of Cardiff.   One and a half miles south-west of Lavernock, lying in close proximity to the shore, is the small island of Sully, a low ledge of grass-grown rock carrying on its summit a small earthwork.   A flat but rocky shore fringes the mainland for three miles between here and the neighbouring Barry Island, and midway, standing back from the sea in a cluster of trees, is the village of Sully, where are the foundations of a castle. At Barry the land again abruptly rises, and the ridge terminates in the cliffs of Barry Island, now insular only in name, for an embankment carrying both a road and a railway connects it with the shore.

A strange transformation has taken place in recent years in Barry and its neighbourhood. From a rural hamlet it has become a flourishing port, for a magnificent series of docks has been made beneath the shelter of the island, which forms a natural breakwater, and a large town has grown up on the hilly slopes behind the docks. The district still possesses some vestiges of antiquity ; on the island are the remains of a small chapel, and on the mainland is the ruined gateway of a castle. The island is indented by a little sandy bay, which provides the public with a bathing place. A small arm of the sea, which once formed the straits, separates Treharne's Point, the most westerly extremity of the island, from Coldknap Point on the mainland, and beyond the latter opens the charming little bay of Porthkerry, floored with pebbles and walled in with limestone cliffs and banks of foliage. On the top of the cliffs is an entrenchment known as the Bulwarks. A wall of lias cliffs, so regular in stratification that it looks like a piece of unfinished masonry, extends without interruption from Porthkerry to the mouth of the Daw, where a cluster of homely-looking cottages forms the village of Aberthaw. An occasional sloop resting on the oozy banks of the estuary and awaiting a cargo of lime is the only reminder that Aberthaw was once a port. On the other side of the stream a wide promontory of low-lying land, called the Lays, spreads as far as Breaksea Point, the most southerly extremity of Glamorgan. Beyond the point Limpert Bay, overlooked by the village of Gileston, divides the promontory from Summerhouse Point, where is another ancient encampment. Here the cliffs again rise and run with but little intermission as far as the estuary of the Ogmore, a distance of 11 miles. Several interesting features, however, occur in the almost unbroken line of cliff. Two miles beyond the Summerhouse, the Colhugh brook finds its way through a narrow vale into the sea. On the summit of the cliff overhanging the left bank of the stream are some more earthworks, known as the Castle Ditches, and a mile up the glen lies the ancient town of Llantwit Major, the site of a famous Celtic monastery. A mile beyond the mouth of the brook are the Tresilian caves, a series of caverns scooped by the waves out of the sides of the cliff and once notorious as a resort for smugglers. Standing on the cliffs beyond the caves is St Donat's Castle, a modernised medieval residence of some repute, and in a glen between the castle and its watch tower is St Donat's church. At St Donat's the cliffs bulge boldly into the sea and terminate westwards in Nash Point, one of the most conspicuous headlands in the Bristol Channel. The point is lighted by a pair of lighthouses, which also serve to warn the mariner off the dangerous Tusker reef beyond.

The coast, which from Lavernock has hitherto kept a

consistent course westwards, here trends north-west as far
as the mouth of the Nedd, but it retains its cliffy character
only until it reaches the estuary of the Ogmore, where
the shore becomes a vast wilderness of sand. The last
noteworthy feature which it exhibits before changing its
aspect is Dunraven Head, which breaks off into the sea
in a sheer wall of rock some 200 feet in height. Behind

Dunraven Bay

the headland lies Dunraven Castle, and close by is the
little watering-place of Southerndown.

Two miles beyond Southerndown the Ogmore river
runs out into a sand-locked bay which may be regarded
as commencing the second section of the Glamorgan
coast, for here the scenery undergoes an abrupt and
striking change. From a series of unbroken cliffs we

pass to a region of unrelieved sand.   The sand indeed is everywhere.   The bay is not only floored with it, but is encircled by a chaos of wind-tossed dunes piled up in places into fantastic pyramids.   Hidden behind the engulfing sand-hills are the village and ancient church of Newton Nottage.   Beyond the bay the underlying rocks emerge for a while from their sandy envelope to form the promontory of Porthcawl Point, on which stands the popular watering-place of Porthcawl.   The town possesses a small dock, but it has now abandoned its trade and given itself up to the entertainment of its summer visitors.   The rocks are low, but line the shore for three miles to Sker Point, where the sand begins again, and forms an arid wilderness all the way to Port Talbot, a distance of six miles.   In the midst of this waste is Kenfig Pool, a large sheet of fresh water around which a few scattered houses mark the site of the now obliterated town of Kenfig.   A gaunt fragment of its castle may still be seen protruding from the sand.

Nine miles beyond Porthcawl the Avon divides the ancient town of Aberavon from its flourishing suburb Port Talbot, which possesses some extensive docks and works.   At Aberavon, which stands on the further side of the river, commences the wide sweep of Swansea Bay. The bay is shut in with high hills and thickly encircled with sands ; and at Briton Ferry the entrance to the Nedd river, which falls into the middle of the bay, has to be protected by long breakwaters.   Briton Ferry possesses a dock, and smaller vessels can go three miles higher up the river to Neath.   Four miles of burrows, which form a fringe to an extensive morass, Crymlyn Bog, reach from the mouth of the Nedd to the estuary of the Tawe, where a dense cloud of smoke indicates the position of Swansea. Opening into the estuary are a number of imposing docks, and the valley of the Tawe has been converted into a vast scene of industrial activity chiefly concerned with copper-smelting. Beyond the Tawe the bay sweeps round in a graceful curve of six miles to the Mumbles Head, its western extremity.

At the Mumbles the Glamorganshire coast enters upon its third phase, for here commence the wild and beautiful cliffs of Gower. The Mumbles Head, an elevated limestone ridge, terminates in a couple of rocky islets, one of which carries a lighthouse and signalling station. At the northern foot of the headland lies the village of Oystermouth, the local centre of the oyster fisheries, where there are a pier and a ruined castle. Its chief attraction is, however, the splendid view of Swansea Bay which it commands. The contour of the Gower coast is so extremely complicated that it is impossible to trace it in any detail. From the Mumbles Head to Worms Head it measures 20 miles, and for the most part it consists of a series of precipitous cliffs indented with a number of sand-strewn bays, all alike beautiful and quite charming in their variety. Between the Mumbles and the next prominent headland, Pwlldu, are Langland and Caswell Bays, both favourite resorts with holiday-makers. Beyond Pwlldu Head opens the extensive bay of Oxwich. Though walled at its extremities with lofty cliffs, which in places exhibit some interesting bone

Worms Head

caverns, the centre of the bay is a sandy hollow overlooked by the ruins of Pennard and Penrice castles, and another ruined mansion clings to the wooded slopes of Oxwich Point. On the further side of this massive promontory is another bay, in which nestles the village of Port Eynon, and from Port Eynon Point a wild and impressive series of cliffs stretch for five miles to the Worms Head, which

Rhossili Bay

forms the western termination of the peninsula. This wall of rock is pierced in places with caverns, the most remarkable of which is known as the Culver Hole. The "Worm," which forms such a fantastic finial to the county, is a long and narrow ledge of limestone, projecting into the sea and ending abruptly in a wedge-shaped crag, some 200 feet high. Its resemblance to a dragon, from which it takes its name, is very remarkable. On the

northern side of Worms Head are the village and bay of
Rhossili. The bay faces west and behind it rises a chain
of breezy downs, which form some of the highest land in
Gower. The northern extremity of this extensive bay is
formed by the Llangenydd Burrows and the little islet
of Burry Holm, whereon is a ruined chapel. Overlooking
the burrows is the camp-crowned hill of Llanmadoc,
whence the coast turns abruptly eastward to form the
Burry Inlet, a shallow and sand-choked estuary which
separates the peninsula from the opposite coast of Carmarthenshire. A long tongue of land, the Whiteford Burrows,
obstructs the mouth of the inlet, and the Glamorgan
shore henceforth subsides into a lengthy and uninteresting
tract of salt-marshes which stretches all the way to the
mouth of the Loughor river. Standing on the ridge of
land which walls in the marsh are the villages of
Llanmadoc, Cheriton, and Llanrhidian, and between the
latter two places are the remains of Landimore and
Weobley castles. At Penclawdd, near the Loughor end
of the estuary, cockles are found in great numbers.

9. Coastal Gains and Losses: Sand-banks and Lighthouses.

In prehistoric days the coast of Glamorganshire must
have been much nearer that of Somerset than at present.
The estuary of the Severn was then but a "silver streak"
dividing Wales from England, and the bed of the Bristol
Channel was an extensive tract of forest land. "There

rolls the deep where grew the tree." Beds of peat and remnants of submerged forests are still to be seen off both the Welsh and English shores. A considerable subsidence took place, which let in the sea and enlarged the estuary of the Severn until it assumed the dimensions of the present Bristol Channel. The original bed of some of the Glamorganshire rivers is 60 feet below the surface of the silt at their mouths. This process of enlargement is still going on. Everywhere the coast is undergoing slow and steady modification. The encroachments of the sea are taking place chiefly at the expense of the Welsh shore. The detached masses of rock which occasionally crop up in the channel were once part of the mainland, and have become isolated from the parent shore by the action of the waves. The beach which fringes the cliffs both of the Vale of Glamorgan and the peninsula of Gower is floored with slabs of rock which once formed the foundations of the cliffs when they stood further out to sea.

The alteration of the coast-line depends, of course, upon the general set of the currents and the character of the shore ; and a glance at the map will show the kind of alteration which we should expect along the coast of Glamorganshire. The general trend of the Channel currents under the stimulus of the prevailing south-west winds is towards the north-east. The peculiar and extensive projection of the Glamorganshire coast exposes it to the full fury of the Atlantic storms. It is evident that under such circumstances it will undergo a constant process of denudation. If land is deposited at all, it will be deposited in those coastal recesses which receive the backwash of the tide, or in those places where rivers can with least hindrance deposit the silt which they bring down from the hills. Fortunately the coast of Glamorganshire is fenced throughout the greater part of its length with a wall of high and fairly hard cliffs, which enables it to resist to a large extent the vigorous assault of the Atlantic breakers. The toughest rocks are those which line the peninsula of Gower. There the wasting of the land is not very considerable. The lias cliffs which border the Vale of Glamorgan are much more easily disintegrated, and from time to time they fall in great masses. Here the erosion is fairly rapid, but is to some extent impeded by the underlying limestone which shows itself in the isolated masses of Sully and Barry Islands. The sea is, however, making the greatest inroads on the shores of Ogmore and Swansea Bays. Almost within living memory the road which connected Swansea with the Mumbles ran along a tract of foreshore now habitually covered with the tide.

On the other hand some compensation is being offered by the sea for this lost land in the shape of various deposits. The most considerable of these gains are the alluvial plains formed at the mouths of rivers. A great deal of valuable land has been created between the estuary of the Rhymney and Penarth Head. This has been made not only by the soil brought down by the rivers, but in still greater measure by the mud washed up by the sea. The fields so won from the tide have to be preserved from inundation by the erection of sea-walls. There is again another tract of low-lying land spread at the mouth of the Daw. Some land has been reclaimed behind Oxwich Bay, though considerable difficulty is still experienced in keeping the tide out, and probably both Crymlyn Bog and the Llanrhidian marsh are only awaiting the enterprise necessary to convert these desolate morasses into valuable pasture land. Far less serviceable than these estuarine deposits are the vast accumulations of

Oxwich Marsh

sand which occur at frequent intervals along the coast. At the mouth of the Ogmore, on the eastern side of Swansea Bay, in Oxwich Bay, and along the shores of the Burry Inlet are, as we have seen, immense quantities of blown sand. These sand-drifts are only the wastage of the rocks, which have been reduced to powder by the ceaseless pounding of the waves. They are deposited by tidal currents and, by cutting off the underlying mud flats from the action of the sea, cause an extension of the land. Their advance is, however, scarcely less destructive than the encroachment of the waves. At Kenfig a fertile land has been turned into a wilderness. The dunes are formed by the persistent and regular action of the south-westerly breezes, which are always carrying the sand further inland. Its forward march can only be arrested by the intervention of vegetation, and the dwellers in these sand-infested localities are under an obligation to plant the marram-grass (*Psamma arenaria*), which has been found most efficacious in stopping the progress of these moving hills. In some places the channel currents sweep along rounded boulders and pebbles, which are occasionally thrown up by strong winds into long banks like the pebble ridge at Coldknap Point.

The sand is as great a source of trouble to the sailor as to the landsman. Owing in a great measure to the numerous sandbanks, the navigation of the Bristol Channel is notoriously dangerous. One-seventh of the total wrecks which take place around the coasts of Britain occur within its waters; and the Welsh shore is chiefly responsible for the fatalities. The fairway to the Channel ports lies much nearer the Welsh side than the English, and the prevailing storms beat down directly upon the Welsh coast. The chief sandbanks are the Cardiff Grounds, a patch of sand lying almost parallel with Penarth Head; the Old 1-fathom Bank, three miles south of Sully Island; the Nash, Middle Nash, and West Nash Banks, a long ridge stretching eight miles west of

Nash Point; the Scarweather Sands, an immense bank three miles west of Porthcawl; the Outer Green Grounds at the entrance of Swansea Bay; the Green Grounds and the Mixon Sands off the Mumbles Head; the East and West Helwick Banks, a long narrow ridge nearly nine miles long, south of Worms Head; the Hooper Sands at the mouth of the Burry Inlet; and the Llanrhidian and Bacas Sands which choke the estuary of the Loughor river. In addition to these sandbanks there are one or two outlying reefs of rock which add to the dangers of the coast. The Wolves are submerged rocks lying between Penarth Head and the Flat Holm, the Tusker is a well-known and dangerous reef at the entrance of Ogmore Bay, and the White Oyster Ledge is a small shelf of rock south of the Mumbles.

The Glamorganshire coast, however, is well lighted, and all the banks and obstructions are indicated by buoys. The chief lights are the Monkstone Beacon, the Flat Holm lighthouse, the Breaksea lightship, the twin light-houses on Nash Point, the Scarweather lightship, the Mumbles lighthouse, and the lighthouse on Whiteford Point. Of these the most conspicuous are the light-houses on the Flat Holm, Nash Point, and Mumbles Head. On the Flat Holm at the south extremity of the island is a white tower 99 feet high carrying an occulting light visible for 18 miles. The two tall Nash lighthouses stand near the edge of the cliff 1000 feet apart, and carry fixed lights which are visible for 19 miles. The Mumbles light is mounted on the outer of the two islets and periodically eclipses. Its flash can be

Mumbles Head

seen for 17 miles. In addition to the lights already enumerated, the harbours of Cardiff, Penarth, Barry, Porthcawl, Port Talbot, Briton Ferry, and Swansea have their own distinguishing beacons, and there are illuminated buoys to mark the fairway for vessels approaching the ports.

Port Eynon Bay

In spite of these precautions, however, maritime disasters are still very frequent on the shores of the Bristol Channel, and the protective work of the Trinity House has to be supplemented by the rescue work of the National Lifeboat Institution. There are lifeboat stations at Penarth, Porthcawl, the Mumbles, and Port Eynon.

10. Climate and Rainfall.

The climate of a country is, briefly, the average weather which it, as a rule, experiences. This depends upon a variety of circumstances of which it is only possible to enumerate the chief. The proverbial uncertainty of the weather shows how multitudinous are the influences which affect it. If it were possible to ascertain them all, the weather could be predicted with absolute precision. The weather in some parts of the world is much more capricious than in others because the circumstances which determine it are less constant and regular. The climate of England is peculiarly fickle because the atmospheric conditions prevailing on these shores are in an habitual state of alteration.

The two chief features of the climate of a country are the temperature and the rainfall, or the amount of sunshine and rain that on an average it receives; and these are determined by two principal considerations, its distance from the equator and from the sea. The sun is the great source of heat, and the nearer a country is to the equator, the more sunshine does it receive, and the higher will be its temperature. The sea, on the other hand, is the great reservoir of moisture, and the closer a country is to the sea, the damper will be the atmosphere, and the greater, in consequence, will be the rainfall. But proximity to the sea likewise affects the temperature. Water is heated more slowly than land, and does not cool so quickly. A country, therefore, which borders the sea

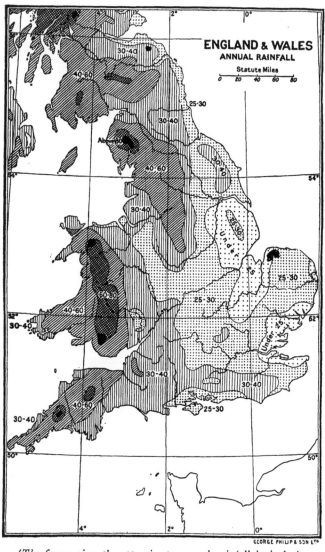

(The figures give the approximate annual rainfall in inches.)

does not experience such varieties of temperature as one further away. In other words, its climate is more equable. The climate of Britain is much more equable than that of countries in the same latitude, for it is not only entirely surrounded by the sea, but owing to the prevalence of south-west winds in the Atlantic the warm waters of the southern regions reach its shores and have a great effect in raising the temperature. So strong is the influence of the north-eastward-moving mass of warm water that the temperature in England varies rather from west to east than from north to south. The ordinary winter of the Shetland Islands hardly differs from that of the Isle of Wight, whilst London is some six degrees colder in winter than Cornwall.

But though latitude and seaboard are the two chief circumstances which determine the climate of a country, they are not the only factors. Amongst other agents which influence it are the altitude of the country, its configuration, its vegetation, and its soil. The higher a land is, the colder it is; partly on account of the chilliness produced by the more rapid evaporation of its moisture and partly on account of its exposure. The highest regions are regions of perpetual snow. A mountainous country is, however, not only a cold country but a wet one, because the clouds are driven up the sides of the hills into the higher and colder regions of the atmosphere, and have a portion of their moisture condensed into rain. Again, the configuration of the land may modify the climate experienced. A range of hills often shelters the plains at its feet from cold winds and frosts. The slopes exposed to the sun not only absorb its heat but radiate it, and so make the surrounding districts warmer. On the other hand, if not so advantageously situated, hills may obstruct the sunlight, and throw a cold shadow over the valleys below. Again, hills not only modify the temperature of a locality, but they considerably influence its rainfall. If they stand close to the seaboard, they chill the moisture-laden air which passes over their summits, and the condensed vapour falls in rain on them and on the districts immediately behind them, so that owing to intervening mountain-barriers lands lying away from the sea are frequently wetter than those on the coast. Vegetation also produces moisture. A fruitful land has sometimes been made a wilderness by being stripped of its trees. Soil likewise affects the climate. A light porous soil makes a country drier and less susceptible to fog than a thick heavy soil. Even the industries of a locality are not without their influence on its climate. The smoke of a manufacturing district will not only obscure the sky and shut out the sunshine, but it impregnates the air with solid particles upon which moisture is readily deposited, and the atmosphere is, in consequence, rendered thick and muggy.

A consideration of these facts will enable us both to determine with some degree of exactness the climatic conditions of Glamorganshire and to account for their peculiarity.

The very extensive seaboard of our county prevents it from experiencing great variations of heat or cold, for the proximity of the sea moderates both the heat of the summer and the cold of the winter. The climate of Glamorgan is not only very temperate but also remarkably equable. The average temperature of England is 48° F., that of Glamorgan is 50°. The Greenwich average is 49°. The mean temperature in the summer is 62°, that of the winter is 38°, which shows only a difference of 24°. The maximum temperature recorded in 1912 was 84·1°, and the minimum 18·8°. So genial are the summers that grapes ripen in the open, and so mild are the winters that myrtles, magnolias, and fuchsias flourish out of doors. But so varied are the physical features of the county that the climate, though equable as a whole, displays remarkable variations in parts. The south-east portions of the county are slightly warmer than the south-west. The hilly district of the north is particularly cold in winter, as the elevation is so considerable, and, roughly, the temperature falls 1° with every 270 feet rise in altitude. The average winter temperature at Cardiff is 37° but at Dowlais only 33°, though Dowlais is only 22 miles north of Cardiff. On the other hand, owing to the moderating influence of the sea, the hilly districts are in summer hotter than the coast, the thermometer often registering as much as 85° Fahr. in the shade. Though its climate is much warmer than that of many other places in the British Isles, Glamorganshire enjoys less sunshine than other counties similarly situated. The smoke of its innumerable furnaces and its frequent sea mists make the atmosphere thick and murky. The sunniest districts in England get on an average 1800 hours of bright sunshine yearly, out of a possible 4450. Cardiff in 1908 received 1669 hours and in 1912 only 1353. Other parts of the county, especially some of the gloomy mining valleys, receive much less.

Let us now turn to the rainfall. Owing to the prevalence of moisture-laden south-west winds from the Atlantic, the climate of the British Isles is particularly humid, and one of the wettest parts of Britain is Wales. The west and south-west districts receive the most rain, as the winds part with a great deal of their moisture as they pass eastwards. The Welsh mountains form a rain screen for central England. The rate at which the rainfall diminishes as we pass from west to east is easily seen from the map. The western counties of Wales have an average rainfall of from 40 to 60 inches and in the central districts of Wales it varies between 60 and 80 inches. In some of the eastern counties of England it averages less than 25 inches. Glamorganshire, owing to its extraordinary diversity of surface, has a very variable rainfall. Though it is drier than many other portions of Wales, its climate is in general considerably wetter than that of England as a whole. The average rainfall of Glamorgan is 50 inches, that of Merionethshire, the wettest part of Wales, is 80 inches (though in particular spots it may be very much higher than this), whilst the general average for England only amounts to about 33 inches. The quantity of rain received by different parts of the county, however, varies very considerably. The wettest district is the mountainous region in the north, and the driest is the sea coast of the Vale of Glamorgan.

The centre of the humid zone is the neighbourhood of Craig-y-Llyn. The Rhondda valley in particular has a bad reputation : it catches all the storms that float over the mountains. The gauge at Treherbert at the top of the valley records an average of 77 inches. This figure is, however, surpassed by the returns from Glyncorrwg on the south slope of Craig-y-Llyn, which yield an average of 83·32. On the other hand the amount of rain on the seaboard of the Vale of Glamorgan is very moderate. At Cadoxton only 32·65 inches fall, and at Fonmon Castle 36·49 inches on an average. In the neighbourhood of Cardiff the figures are rather higher. It will be noticed that the rainfall bears a fairly constant relation to the altitude, the difference in height between the places giving the highest and lowest returns being 697 feet. A curious exception to this rule is, however, furnished by the returns from Cwm Bargoed overlooking the Rhymney valley, where the altitude is 1225 feet. Here the fall recorded in 1912 was 78·91 inches, and the average is given as 61·88, figures which are scarcely higher than those obtained in other places of little more than half the altitude. But this is the exception which proves the rule.

Glamorganshire, unfortunately, does not possess the clear skies and transparent atmosphere of less busy regions. The general humidity of the climate and the prevalence of smoke make the air as a rule hazy. Fogs are of frequent occurrence in the Bristol Channel, but they are generally of a drifting character. Ground fogs are not prevalent except in the low-lying neighbourhood of Cardiff and in the vale of Neath. The hills are, however, frequently wreathed in mist, and the atmosphere of the mining valleys is exceedingly murky, for the altitude of the adjoining mountains prevents the smoke from dispersing.

11. People—Race. Language. Population.

We generally speak of the Ancient Britons as the aboriginal inhabitants of this island, and we regard the Welsh as the chief survivals of the stock. As a matter of fact Wales is no more the home of one ancient people than it is of one ancient language. Many races have reached its shores, and have contended amongst themselves for the possession of its soil. Some have perished and some still survive ; but nearly all have left memorials of their sojourn amongst its vales and hills. The earliest traces of human occupancy are the flint implements of a race of rude hunters who shared with the beasts of the field which they pursued the scanty shelter of the cave. It is not, however, to these that we must look for the early progenitors of the Welsh people. The Welshman of to-day is the offspring of two distinct races of immigrants who pressed westwards from different parts of the continent.

The first comer was the Iberian, short, swarthy and long-headed, who arrived possibly from the northern shores of the Mediterranean. The Iberian was succeeded by the Celt, who came from some colder home on the

plains of northern Europe.   The Celt was tall and fair,
and with weapons of bronze and iron he dispossessed and
enslaved the simpler and less instructed Iberian armed
only with weapons of flint.   The Celts came in two
successive bands, the Goidels or Gaels and the Brythons,
whose arrival was separated by a wide interval of time.
The Goidels arrived at a very early period—perhaps
1200 B.C.—and brought with them weapons of bronze.
They swept over Britain and Wales, and spread even to
the shores of Ireland.   It was not until the fourth century B.C. that the Brythons made their appearance, and
when they came, they came armed with implements of
iron.   Their conquest of Britain was sufficiently complete to give to its people the familiar name of Britons.
How far they succeeded in occupying Wales is a moot
point.   Probably for a time they penetrated only into the
central parts of the Principality, and did not overrun
the country until after the departure of the Romans in
the fifth century A.D.   But it is not to either Goidel or
Brython that the Welsh owe all their peculiarities.   The
Welshman of Glamorgan, at any rate, is not purely and
entirely a Celt.   All through the contest between Gael
and Briton the Iberian survived, living side by side with
his successive Celtic masters.   To-day two types of
people exist, the one short and dark, the other tall and
fair, each of whom may be regarded as perpetuating the
original characteristics of the two different races.   Everywhere in Wales these two types are to be met with, but
nowhere is the dark skin of the Iberian more plainly
exhibited than in the county of Glamorgan.   To-day the

dark type is the predominating element in the population, and it is to his Iberic blood that the native no doubt owes his imaginative and emotional nature.

Glamorganshire has, however, become the home of still another people. Besides the Welshman, there is the Englishman. Iberian, Goidel, and Brython in course of time forgot their racial differences, and coalesced into one nation. They became Cymry—" fellow countrymen." But the Englishman has always been regarded as an alien. This is due in some measure to the fact that he came originally in the train of the Norman conqueror. The English district is the district which was overrun by the Norman nobles, whose followers settled down on the land which they annexed. The inhabitants of the Vale of Glamorgan, which corresponds with the Norman sphere of influence, are much less Welsh than their neighbours in the hills. The district round Llantwit is peculiarly English. The people of Gower are not only English in race but English in speech. Some have found an explanation of the foreign character of these districts in a tradition that Henry I planted colonies of Flemings both at Llantwit and in Gower. The tradition, however, lacks historical confirmation. If the Flemings came to these parts at all, they probably came in the train of the Conqueror's Flemish lieutenants, some of whom settled in Somerset and sallied forth from their strongholds on the banks of the Parret to seek further adventure in Wales. The maritime position of these localities is in part an explanation of their peculiarity. Since the Norman opened up Wales to the Englishman, there has

been a constant commerce between the Welsh and
English shores of the channel, and there seems some
reason to believe that Gower at any rate has been largely
peopled from Somerset. The change in the industrial
character of the county generally has still further leavened
the population. The inhabitants of the sea-ports are now
largely cosmopolitan. In the mining districts also there
has been a considerable influx of outsiders, not only from
England but from Scotland and Ireland.

If there is some difficulty in tracing the pedigree of
the Welshman, there is none in tracing the source of his
language. The Brythonic invasion, though it did not
exterminate the Iberian and the Goidel, nevertheless completely eradicated their languages. The Welshman, unlike
his kinsman in Scotland and Ireland, does not use the
Gaelic but the Brythonic tongue. The only trace of his
mixed ancestry in his speech is his method of arranging
his words—a peculiarity which seems to indicate the
lingering influence of earlier idioms. In one respect the
Celt is debtor to the Roman. The Roman occupation
was merely an historical interlude, but it lasted long
enough for the Welsh tribesman not only to borrow from
the Roman soldier certain military and domestic terms
but to acquire the art of writing. Before the Roman
occupation the Celt appears to have had no method of
transcribing his language beyond the rude Ogam signs.
When at last he learnt to write he wrote in Roman
characters.

The complete disappearance of Gaelic left Welsh the
universal language of Wales. Now, however, in South

Wales, English is proving a formidable rival. Everyone in Glamorganshire can speak English, though for sentimental reasons Welsh is still largely adhered to. It is computed that 35 per cent. of the population use Welsh as the habitual vehicle of conversation. The Welsh-speaking districts are chiefly in the north and west. In Cardiff the Welsh-speaking population scarcely numbers 10 per cent. of the inhabitants, whereas in the neighbourhood of Swansea half the people, it is said, habitually use Welsh. The dialect of the county, owing probably to the intermixture of races and the persistence of Iberic influence, possesses many peculiarities. It is especially soft in character, and its vocabulary retains many archaic words. English is universally employed in the peninsula of Gower, but though it bears a certain resemblance in phraseology to the West Saxon speech of Somerset, it is spoken with a Welsh accent and intonation.

Though the population of Glamorganshire does not rival in number that of some of the large industrial counties in the north of England, yet nowhere has growth been more rapid. In a century its inhabitants increased more than ten-fold. In 1801 the estimated population was 70,879. In 1901 it was recorded as 859,931. Now it has reached 1,121,062. In the last decade its people have multiplied more rapidly than those of any county in England and Wales outside the metropolitan area. Glamorganshire now contains more than half the total population of Wales. Compared with the average population of England and Wales there are 1382 people to every square mile as against 669 in England

and 271 in Wales. In other words there are twice as many people to a square mile in Glamorganshire as there are in England, and five times as many as in Wales. There are now in Glamorganshire nearly two people to every acre of ground. Contrasting these returns with those of the individual counties at the extreme ends of the scale, the population of Glamorganshire is about seventeen times as dense as that of Westmorland, the most thinly inhabited of all the English counties, and twenty-eight times as dense as that of Radnorshire, which stands at the bottom of the list of Welsh counties. The population is very unevenly distributed. The three largest towns, Cardiff, Swansea, and Merthyr, contain together more than a third of the total population of the county. In recent years there has been a large shifting of the population. People have flocked from the country into the towns and mining districts. Both the Vale of Glamorgan and the peninsula of Gower have been depleted of their inhabitants. On the other hand many once rural districts have, owing to mining operations, become densely populated. The Rhondda valleys, for instance, formerly one of the most sparsely populated districts in the county, now have over 152,000 inhabitants and are, after Cardiff, the largest urban area in Wales.

12. Agriculture.

Though Wales is largely an agricultural district, its exceedingly mountainous character makes it compare unfavourably with England in its agricultural returns.

In England 24½ million acres are in cultivation out of
a total of 32 millions. Mountainous land possible for
grazing purposes, but unsuitable for other agricultural
uses, amounts to two million acres. In Wales the
mountainous land reaches a much higher proportion.
In Glamorganshire the mountain and heath land useless

A Glamorganshire Farm

for the purposes of the farmer covers 132,818 acres.
Out of a total area of 518,865 acres only 261,363 acres
—little more than half—are under cultivation. Of these,
51,847 acres are arable land, and 209,516 are laid down
in permanent grass, of which 74,067 acres is hay grass,
and 125,449 pasture. Woodland accounts for 25,712

acres. The proportion of pasture land is generally tending to increase. The cereals chiefly grown are oats, wheat, and barley. Of these oats take 10,635 acres, wheat 4296 acres, and barley 5576 acres. Rye is but very slightly cultivated, only a few acres being grown. Of the other crops raised, potatoes cover 1620 acres, turnips 6185 acres, and mangolds 1397, while of clover and other fodder there are 20,169 acres. Of land devoted specially to the production of fruit, 264 acres are employed in the cultivation of marketable varieties, of which apples, accounting for 231 acres, are the most important feature. Only small quantities of strawberries, currants, gooseberries, plums, and cherries are grown. Vineyards have been tried with fair success at Penarth and on the sunny slopes near Castle Coch, and experimental crops of tobacco have been produced in the neighbourhood of Cardiff. Of the total amount of land under crops of various kinds 14,371 acres are cultivated by the owners and 246,992 by tenants.

Though the more fertile parts of Glamorganshire are very suitable for the production of corn, the farmer has found it more profitable to turn his fields into grass. The immense increase in the size of the towns has given rise to an enormous demand for milk and dairy produce, and the facilities which the ports afford for the importation of cheap feeding-stuffs have led the agriculturist to prefer the rearing of cattle to the cultivation of wheat. Large numbers of cattle are therefore bred upon the pastures. Glamorganshire in days gone by produced a breed of cattle of its own, but this distinctive variety has now almost disappeared. There is now no uniformity in the kind of cattle fed in the county. The herds are extremely miscellaneous, Herefords, Devons, and shorthorns are met with indiscriminately. The abundant rainfall and warm climate make the herbage rich and luxuriant, and the cattle reared on the Glamorganshire pastures compete on equal terms for the favour of the butcher with the best west of England beasts. Butter is made in large quantities and cheese is manufactured to some extent. At Caerphilly a soft white cheese is produced which was once peculiar to the county, but it is now manufactured under the same name on the other side of the Channel. The number of cattle in the county is given as 56,071, and the horses amount to 17,806.

The mountain and moorland are used very largely for sheep rearing. There are now 319,607 sheep within the confines of the county. Those fed on the mountains are chiefly a small nimble-footed kind which is of little use to the wool-merchant, though their mutton is highly esteemed. On the plain the flocks are, like the cattle, very varied in breed. Leicesters and Downs perhaps predominate, but almost every variety, long and short woolled, white-faced and black, may be met with. Pigs are reared to some small extent, chiefly by cottagers, but they are not bred in such quantities as formerly. They now amount to 17,915.

Glamorganshire farms are as a rule of considerable size, but agriculture, generally speaking, is somewhat backward, and farmers are inclined to adhere to the older methods. Lime, which is a natural constituent of the soil, is much employed, large quantities of limestone being burnt, and the lime spread on the fields.

13. Industries and Manufactures.

The industries of Glamorgan, if not of recent introduction, are all of recent growth, and their development has followed closely on the opening-up of the coalfield. The prosperity of the county depends on its stock of coal. The exhaustion of the coal-supply would be followed at once by a complete commercial collapse. The rise of the county into the front rank of manufacturing centres has been the work of the last hundred and fifty years. A century and a half ago Glamorganshire was known chiefly as an agricultural district. Beyond a little coal-mining, copper-smelting, and " hardware " manufacture, no other industry existed. To-day its volume of trade is surpassed by few counties in the kingdom. In spite, however, of the phenomenal amount of business done, its industries are of a rather restricted range. Apart from the coal and metal trades and their allied businesses, the manufactures are few, and consist chiefly of brick and tile making, pottery manufacture, chemical works, rubber works, brewing, and paper and flour mills. The enterprises in connection with the coal and metal trades are, on the contrary, most varied and extensive. Every branch and ramification of these industries is represented. Iron, copper, nickel, silver, and zinc smelting, the manufacture of steel rails, bolts, and bars, steel wire and tubes, armour-plates, chains, anchors, steam-engines, rolling stock, winding and hauling machinery, colliery plant, galvanized sheets, tin and terne plates, patent fuel and fire-bricks, are all carried on.

The early history of iron smelting in the county is obscure. The beds of iron cinders which have been discovered at Ely and St Nicholas seem to show that the Romans were not altogether unacquainted with the mineral resources of Glamorgan. In the former district a small foundry hearth has been unearthed on the site of a Roman building, but there were no indications that the furnace had been used for producing iron except for domestic purposes. Of the local manufacture of iron in medieval days we know little. The existence of "bloomeries," or old smelting-hearths worked by hand-bellows, shows that iron must have been made to some extent by the inhabitants. Leland speaks of ironworks existing in the neighbourhood of Llantrissant in 1539. Furnaces are said to have been in operation at Aberdare in the days of Edward VI, and bars of old iron discovered amongst local ash-heaps seem to confirm the tradition. In the reign of Elizabeth two iron furnaces are recorded as having been set up at Radyr in the Taff valley by Sir W. Mathew, and a Government Order of 1602 prohibits one of his descendants from "casting ordnance at his furnace near Cardiff" lest the facilities of transport offered by the port might tempt him to supply the Spaniards with guns. In 1666 another furnace was in full swing at Hirwain, and in 1680 John Morgan and Roger Powell established a smelting-hearth at Caerphilly. But if the industry existed, it can only have been on a small scale, for as late as 1740 the total output of iron from Glamorganshire was only 400 tons a year. It was not until the middle of the eighteenth century that the industry really commenced in South Wales. Its pioneers were Anthony Bacon and John Guest. In 1758 a Mr Lewis of the Van, who had already started a small furnace at Pentyrch, secured the services of John Guest, a Staffordshire man, to open new works at Dowlais. The undertaking grew into the famous Dowlais ironworks, which is now one of the leading industries in the country. In 1763 Anthony Bacon, who was M.P. for Aylesbury, began the Plymouth and Cyfarthfa ironworks, and in 1782 in conjunction with Samuel Homfray he transferred his furnaces to Penydarren. The Cyfarthfa works were subsequently purchased by Richard Crawshay, "the iron king," who so extended the undertaking that by the beginning of the nineteenth century these works alone were employing over 1000 men. It was to the Plymouth and Cyfarthfa works that the Government turned for cannon during the early stages of the American War of Independence.

The first real impetus to the development of the Glamorganshire iron trade came from the discovery of the suitability of coal for smelting purposes. It had been customary to extract iron from the ore entirely by the agency of charcoal, which was obtained by stripping the local woods. It was the substitution of coal for charcoal which opened the new industrial era for South Wales. The success of the early ironmasters stimulated the foundation of other enterprises, and by the middle of the nineteenth century Glamorganshire iron was as famous as the products of Staffordshire. The extreme cheapness with which steel can now be produced has led to its almost universal adoption in the place of iron for manu-facturing purposes. Nearly all the large ironworks are fitted with "converting" furnaces for the production of steel. The annual output of iron now amounts to 500,000 tons, and the output of steel to about the same quantity. The number of iron and steel furnaces in the county is 20, and they employ between them 15,000 men. The chief centres of the iron and steel industry are Briton Ferry, Cardiff, Clydach, Dowlais, Gowerton, Landore, Melincryddan, Merthyr, Pontardawe, Port Talbot, Port Tennant, Swansea, and Ystalyfera.

Copper-smelting, like the iron trade, made an early commencement in Glamorgan. In 1564 a London com-pany, which already possessed copper-works in Cornwall, extended their activities to the other side of the Channel, and erected a "meltinge-house" at Neath, where the industry was carried on by Dutch and German workmen. In 1717 a Dr Lane established a smelting furnace at Landore, and that locality has now become the metro-polis of the trade. The amount of copper ore im-ported into the port of Swansea exceeds 200,000 tons annually, and the quantity of metal produced is over 21,000 tons. The chief copper-smelting furnaces are at Briton Ferry, Cardiff, Cwmavon, Landore, Llansamlet, Morriston, Neath, Port Talbot, Skewen, Swansea, and Taibach. Glamorganshire practically monopolises the manufacture of zinc. Swansea alone produces nineteen-twentieths of the total output of this metal in the United Kingdom. Gold, silver, nickel, and lead ores are also largely smelted in the same neighbourhood. Swansea has been called the "metallurgical capital of the world." Within four miles of the town there are 140 works engaged in smelting operations of one sort or another. No less than 36 different kinds of metallic ores are there treated, and the number of men employed in the industry exceeds 30,000.

Although the tin mines of Cornwall were some of the best known sources of the mineral wealth of England and were worked from the earliest times, yet it was not until the seventeenth century that Englishmen acquired the art of tin-plate making. It was apparently introduced into England from the continent by Andrew Yarranton, a Parliamentary soldier. The industry was first taken up in Pontypool in Monmouthshire, and from thence soon spread into South Wales; but it was not till coal began to be used in the manufacture of iron that it exhibited any considerable signs of development. In Glamorganshire it has now been carried on for more than a century. Unfortunately the trade is subject to violent fluctuations, but in good seasons it reaches enormous proportions. In the Swansea district alone (which is the chief seat of the industry) the output frequently amounts to six million boxes. The principal centres of manufacture are Aberavon, Aberdulais, Briton Ferry, Clydach, Gorseinon, Landore, Llansamlet, Llantrissant, Melincryddan, Melingruffydd, Morriston, Neath, Pontardulais, Skewen, Taibach, and

Briton Ferry

Ystalyfera. The trade in galvanized sheets is nothing like so extensive. Swansea, which exports over 400,000 tons of tin, terne, and black plates, only exports a little over 40,000 tons of galvanized sheets.

An enormous business is now done in the county in the manufacture of patent fuel. A process for utilising small coal by grinding it up with pitch and pressing it into blocks was patented by a Mr Wood in the middle of the nineteenth century. The process has since been considerably improved, and huge quantities of this artificial fuel are now shipped for consumption abroad. The total quantity exported from the various Glamorganshire ports amounts to well over a million and a quarter tons.

The timber trade is another very important local industry. Vast quantities of pitwood are required in the collieries for roof props, and there is a very large demand for sawn timber and boards for general purposes. All the larger docks are furnished with extensive timber floats, and saw-mills are numerous. The importation of pitwood amounts to a million tons and that of ordinary timber to a quarter of a million tons.

Engineering in all its varied branches—mining, marine, and general—is carried on extensively in the neighbourhoods of Cardiff, Neath, Pontypridd, Swansea, and Tondu. At Pontypridd there is a large chain and cable works, and at Cardiff a Lloyd's Proving House. There is a certain amount of sail-making done at Cardiff, but the substitution of the steamer for the sailing-vessel has led to a great decline in this industry. Brattice cloth for use in collieries is manufactured also at Cardiff. At

Penarth and Rhoose there are large works for the production of Portland cement, which consists of a combination of lime and silica. The hard nature of the calcareous rocks in Glamorganshire seems to prevent the general extension of the industry; but sufficiently suitable material is found for its production in the blue lias rocks of the Vale. Aberthaw at one time had a wide reputation for furnishing a hydraulic lime which naturally possessed the essential qualities of cement. Brick and tile making is carried on at Bridgend, Cardiff, Gowerton, Merthyr, Neath, Penarth, Pencoed, and Swansea.

Glamorganshire once made a name for itself in the world of art by the manufacture of a very superior kind of porcelain, but the business has now died out. The first kilns were started at Swansea in 1750 for the manufacture of earthenware. After a preliminary effort to produce "opaque china," an attempt was made in 1814 to manufacture real porcelain. In the same year a rival factory, which had been established by Billingsley at Nantgarw in the Taff valley, removed also to Swansea, but in 1817 returned to its original home. Porcelain continued to be produced there until 1822. The Swansea kiln prolonged its operations for two years later, when it reverted once more to the manufacture of earthenware. The Glamorganshire porcelain, which is still much sought after by connoisseurs, is described as "possessing a translucent body equal, if not superior, to that of the finest old Sèvres." A coarse kind of drab and white ware is still manufactured at an old-established kiln at Ewenny, and a red and black pottery is made at Swansea. Architectural and decorative terra-cotta is produced in large quantities at Pencoed.

Brewing is extensively carried on in Aberdare, Bridgend, Cardiff, Merthyr, Neath, Pontypridd, Swansea, and other places.   Paper mills exist at Ely, vitriol is made at Gorseinon, and at Cardiff there is a huge flour-mill and biscuit factory.

14. Mines and Minerals.

It seems unlikely that the Romans, who valued Britain chiefly for its minerals, and who made good use of the lead and tin mines on the opposite shores of the Bristol Channel, were unacquainted with the mineral resources of South Wales.   There are the remains of some ancient zinc and lead mines at Merthyr Mawr which may be of Roman origin.   It was probably the unsettled state of the country which prevented the Romans from developing to any extent the minerals of Glamorgan.   There is some mention of iron mining in the laws of Hywel Dda, the great Welsh legislator of the tenth century, which shows that mining traditions were current in the district after the departure of the Romans.   In medieval times Glamorgan was known to possess both iron and coal, but the easy terms on which mining rights were let by the Norman landowners suggest that they had no suspicion of the riches within their reach.   The Normans themselves had too much fighting to do to find time for commercial undertakings, and mining, like agriculture, was left for development to the monks. The abbey of Margam seems to have been in quite early days concerned in the mining industry in Glamorgan. Early in the thirteenth century the monks obtained for paltry acknowledgments the right to search for iron, lead, and coal upon the lands of neighbouring landowners, and the abbey amongst

Oakwood Pit, Maesteg
(*One of the oldest in the county*)

other property at the Dissolution is recorded as possessing a coal mine at Cefn Cribwr. There is also in existence at Swansea a fourteenth century charter by which William de Breose empowers one of his tenants to dig "pit-coal" out of his estates, and in 1547 William Herbert secured authority from Edward VI to work the iron ore obtainable in the neighbourhood of Llantrissant. Little is known, however, of the use which either monks or private individuals made of these mining rights.

Though, as already stated, ironworks appear to have existed in Glamorgan in the sixteenth and seventeenth centuries, there is little to show what methods were employed for obtaining the ore. Probably both iron and coal mining were carried on but fitfully until the middle of the eighteenth century. The first attempt to develop the mineral resources of the county in anything like a systematic fashion or on an adequate scale was made by one of the Morgans of Ruperra, who in 1748 purchased as a speculation the mineral rights of the manor of Senghenydd and sublet his interest in them to enterprising men like Lewis of the Van and Anthony Bacon. These facilities led to the establishment of the works at Cyfarthfa and Dowlais. The ironstone was already at hand. It occurred chiefly in the form of yellow limonite in pockets at irregular intervals along the belt of Mountain Limestone which girdles the coalfield. The ore was secured by tunnelling the limestone, which itself served as a flux in the subsequent smelting operations. It was only when charcoal began to get scarce and the suitability of coal as a substitute was discovered that any attention was paid to the coalfield. The coal was secured by means of pick and shovel wherever it cropped out at the surface. Whenever faults or other difficulties were met with, the workings were abandoned and a new seam opened elsewhere.

The demand for coal which was originally created by the iron trade was further stimulated by the invention of the steam-engine, and the need for larger supplies of domestic fuel. The only thing which checked the more rapid development of the coal-mining industry was the lack of adequate means of transport. The construction of the Glamorganshire Canal in 1795 and the subsequent development of the railway made it possible to distribute the products both of the iron furnace and the coal mine to distant parts. In 1840 the first cargo of coal was shipped at Cardiff. The constantly increasing demand for fuel soon made more efficient methods of mining necessary, and in 1850 the first pit was sunk in the neighbourhood of Aberdare. In 1865 the Ogmore valley was opened to the miner, and very soon every valley in Glamorganshire was being energetically exploited. In 1910 there were 408 collieries within the county, and the men employed numbered 134,712. The total production of the whole of the South Wales coalfield in that year reached 48,700,000 tons. In 1912 the amount brought to the surface was 50,116,264 tons, out of a total of 260,567,552 tons raised in the United Kingdom. About 260 tons of coal are raised for every person employed. It is estimated that, notwithstanding the enormous output, there still remain in the Welsh coalfield 28,000,000,000 tons of workable coal, which at the present rate of production (50 million tons) will last 500 years.

Whilst coal mining has so enormously increased, iron mining has, on the other hand, considerably declined.

Numerous seams of clay-band ironstone occur in the lowest beds of the Coal Measures and were once extensively worked for smelting purposes, but the importation of cheap foreign ores of greater purity has led to the disuse of the native material.   Scarcely 10,000 tons of native ore are now obtained annually, whereas 668,482 tons of foreign ore were imported into Cardiff, and 161,936 tons into Swansea in 1910.

Coal Trains on their way to the Docks

Lead mining, though apparently begun by the Romans, has never prospered in Glamorganshire. Small veins of galena can be detected in several places, and attempts have been made to work them in the neighbourhood of Draethen, Cowbridge, and Brocastle, but without success.   There is a small deposit of gypsum in the district of Penarth.   An effort was made to work it between Penarth and Lavernock, but owing to the softness of the overlying strata the venture proved unremunerative.

There are several quarries in the county. The stone chiefly used for building is the Pennant Grit from the upper and middle Coal Measures. It is worked chiefly at Pontypridd, Treforest, Llanbradach, Nelson, and Quaker's Yard. It is a fine-grained and hard calcareous sandstone of a blue colour, but soon changes to yellow owing to the oxidization of its iron compounds. The Triassic rocks in the neighbourhood of Bridgend once yielded a good even-grained sandstone known as "Quarella." The Silurian sandstones obtainable near Cardiff are employed for walling and rough building.

Limestones are much used. Those from the Lias beds of the Vale of Glamorgan, owing to intervening bands of shale, are easily quarried and are fairly uniform in thickness, but they weather badly. The variety known as "Sutton Stone" has proved more serviceable. The Radyr stone—a coarse limestone breccia, occurring at Cadoxton, Radyr, and near Porthcawl—once found favour with builders for ornamental purposes. The Carboniferous limestone is not generally a building stone on account of its hardness and the difficulty of dressing it. It was, however, employed very largely in the construction of the Barry docks, and has proved itself the best road-metal in the district. It is quarried at Wenvoe, Walnut Tree, Thornhill, Hirwain, Cowbridge, and Barry. The Old Red Sandstone is but little used, as it is too soft; and the Millstone Grit is very sparsely employed. The

Lias limestone is quarried for lime-burning and cement-making at Penarth, Rhoose, Bridgend, and Pyle.

Large quantities of fire-clay are obtained from the coalfield, and a number of collieries near the southern outcrop have brickyards connected with them to utilise the clay obtained in the workings. Another kind of

Limestone Quarry, near Porthcawl

fire-brick which is much esteemed is made by crushing the Millstone Grit. For ordinary brick-making the red Triassic clays from the neighbourhood of Roath, Penarth, Cogan, and Llandough are extensively used. The clays of the Old Red Sandstone near Llanishen, and the alluvial clays of the Cardiff moors, also furnish fairly satisfactory brickfields. The clay used in pottery-making is chiefly obtained from the bands found interspersed between the limestones of the Lias formation.

15. Fisheries and Fishing Stations.

Probably on account of its proximity to the Atlantic, whence its stock of fish is being constantly replenished, the Bristol Channel is a valuable and prolific fishing ground. So considerable is the supply that not only are there several Welsh fishing-fleets, but trawlers from Hull and Brixham find it worth their while to come into Welsh waters. The visits of these strangers would probably be even more frequent if the Channel provided a more varied list of marketable fish to tempt them from their own grounds. The chief fishing-stations within the county are Cardiff, Swansea, and the Mumbles.

The methods employed in the capture of the fish vary according to the different species, but are usually either trawling, surface fishing, or dredging. Trawling, which accounts for two-thirds of the fish captured, goes on all the year round, and is employed for the capture of soles, hake, skate, ray, haddock, bream, ling, cod, conger eel, whiting, and pollack. The vessels used are steam-boats of about 70 tons register, manned by nine hands. Cardiff possesses a fleet of 17 of these trawlers, and Swansea 30. Neither of the ports possesses sailing boats; but from April to September a number of Brixham sailing trawlers, each worked by a crew of five hands, make Swansea their headquarters. The quantity of fish landed during 1912 at Cardiff was 141,007 cwts., and at Swansea 174,231 cwts., the value of the respective hauls being £83,265 and £130,982. The total quantity of fish caught off the coasts of England and Wales during 1912 amounted to 14,612,000 cwt., representing a value of nearly £8,900,000, of which one-third was taken by trawlers.

The trawling grounds extend over an area from 20 to 50 miles west of Lundy Island, though the boats sometimes pursue the fish as far as the coast of Ireland, the Bay of Biscay, and the shores of Morocco. The "trawl" used in catching the fish is a net attached to a heavy beam of from 42 to 50 feet in width. The beam is secured at each end to a wire rope, which is worked from a steam capstan, and by this means the net is dragged along the bottom of the sea. Each net is furnished in front of its mouth with a thick "foot-rope" which acts as a stirrer to disturb the soles and other flat fish which lie basking in the sand. Trawling is only feasible in sandy areas.

The most valuable and plentiful of these bottom fishes is the sole. The Bristol Channel has been described as "the home of the sole." A single trawler will in good seasons secure a catch of from 8 to 14 cwt. in a three days' cruise. Its season is from December to the beginning of May. Cod also spawn freely off the Glamorganshire coast. Whiting, hake, and conger eel are abundant in the Bristol Channel, but lemon soles, plaice, and haddock are scarce, and the halibut is seldom met with.

Surface fishing is chiefly done by means of "drift nets"

and "seines," which are mostly handled by the crews of
sailing boats. They are large nets weighted at the bottom,
and buoyed upright in the water by floats. Drift fishing
is usually pursued at night, as the darkness hides the net
from the fish. The species most generally secured by
the drift net are mackerel and herrings, which swim about
in shoals. Mackerel are at times very plentiful, and a
single boat will sometimes catch as many as 30 or 40
thousand in a night. The total quantity of mackerel
captured by the Glamorganshire fleet in 1909 was valued
at £1298. The coast of South Wales seems to be one
of their favourite breeding grounds. The herring which,
either fresh or dried, forms such a universal article of diet
is also very common in the Channel. Though generally
supposed to be an East Coast fish, it reaches its highest
perfection in the west. It is plentiful in the winter, but
the finest fish are secured in the summer—a circumstance
probably due to some change in the character of its food.
The sprat also haunts the Channel, attracted possibly by
the abundance of young shrimps. They appear chiefly
from October to December, and are invariably pursued
by numbers of whiting, which feed on them freely.

The lamprey, from a surfeit of which Henry I is said
to have died, is largely caught in the higher reaches of
the Channel. They make their way up the freshwater
streams for the purpose of spawning.

Another method of capturing surface fishes is by
"whiffing," which consists in dragging an unweighted
net after a row-boat; but the method is only effective in
the case of fish, like the pollack, which occasionally

frequent the very top of the water. "Stop and stake" fishing is occasionally resorted to at the Mumbles and elsewhere where a long stretch of fore-shore is exposed by the receding tide. Nets are suspended from poles placed at intervals along the beach, and the fisherman secures whatever the tide happens to leave behind. This haphazard process is looked upon with great disfavour by the trawler, who believes that it destroys vast numbers of immature fish which frequent the bays and creeks at spawning time. Hook and line fishing is freely indulged in, chiefly for the sake of the sport which it affords. Cod, halibut, pollack, and mackerel are easily caught in this way.

The Mumbles is the chief station for oyster fishing, and the fisher-folk there devote themselves almost exclusively to that business. The Mumbles dredging-fleet consists of about 20 sailing smacks, carrying three hands apiece. The amount of other fishing done by these vessels is small. Out of the £1760 worth of fish landed here in 1909 only £26 worth was put down as "wet" fish. The boats go out to scour the deeper waters of Swansea Bay by means of dredges. The oysters are scooped up and then placed on beds close to shore to mature and await the demands of the market.

Of the fish which divide their time between the sea and the rivers the most valuable is the salmon. The total value of the British salmon fisheries is about £800,000. The salmon is really a salt-water fish which goes up the rivers and estuaries to spawn. The impurities with which the Glamorganshire rivers are impregnated have now practically closed them to the ascent of the salmon and sewin.

The eel, unlike the salmon, is naturally a freshwater fish, but leaves the rivers and seeks the sea at spawning time. Immense numbers of young eels, or elvers as they are called, ascend the tidal rivers and are netted as they go up, whilst the mature fish are caught in the estuaries as they go down.

16. Shipping and Trade. Chief Ports.

Though there are six large shipping centres in Glamorganshire—Cardiff, Penarth, Barry, Port Talbot, Briton Ferry, and Swansea—there are technically only three ports—Cardiff, Port Talbot, and Swansea. A "port" in strictness is a Customs port, and its limits are defined by the Treasury for convenience in collecting revenue. Though quite distinct ports in the ordinary sense of the word, Penarth and Barry are for statistical purposes grouped together with Cardiff, whilst Briton Ferry is grouped with Port Talbot. Before 1846 there was only one officially recognised port in South Wales, the limits of which extended from Chepstow to Llanelly. It seems to have been called the port of Cardiff or the port of Swansea, according to the relative importance of the two places at the time. All the other harbours on the Welsh side of the Channel were "sub-ports" or "creeks."

Notwithstanding the fact that the rise of the Glamorganshire ports was one of the outstanding features of

the commercial history of the nineteenth century, both
Cardiff and Swansea have behind them a very venerable,
if a not too respectable, record as sea-faring places.   As
early as 1326 Cardiff was created for the time being a
"Market of the Staple," and during the next century it
continued to be a place of both export and import.   In
the sixteenth century, however, it appears to have become
a mere resort of pirates.   The ill repute of Cardiff possibly
accounted for the rise of Swansea, for in the reign of
Elizabeth the port of Swansea contributed several ships
to fight the Spanish Armada.   There is, however, a
suspicion that the vessels fitted out at Swansea instead
of scouring the Spanish main, may have helped to swell
the fleet of illicit traders which ransacked the Bristol
Channel from the port of Cardiff.   Swansea, in any case,
seems to have kept the lead which it had obtained, for in
the eighteenth century, when Cardiff had sunk to the
position of a "creek" of Bristol, its rival was doing a
very considerable trade.   In 1768 the vessels cleared
from the port numbered 694, and the registered tonnage
was recorded as 30,631.   The prospects of Cardiff, on
the other hand, were regarded as hopeless, especially in
that particular in which it has now become so famous.
In 1775 the official Customs report stated that "no coals
are exported from Cardiff nor ever can be, its distance
from the water rendering it too expensive for any such
sale."   Nothing perhaps shows more strikingly the magnitude of the industrial revolution brought about by the
development of the Glamorganshire coalfield than a com-
parison between the position of Cardiff then and now.

In little more than a century Cardiff has become the
largest coal-exporting port in the world, the first port in
the United Kingdom for the volume of its foreign
exports, and the third in point of tonnage cleared.
Swansea can exhibit almost equal signs of progress, and
Barry and Port Talbot have been entirely created.

The business done at the Glamorganshire docks is

Cardiff Docks

almost entirely in the form of exports, of which coal
forms the bulk.   Except for timber, grain, potatoes, and
metallic ores the import trade is comparatively insig-
nificant, and does not equal even in the aggregate that
done at Bristol.   Foreign cattle, however, are brought in
large numbers to Cardiff, which has special accommodation
for their slaughter and storage.

The foundation of Cardiff's prosperity was laid by

the second Marquis of Bute, who in 1839 spent a fortune of £350,000 in the construction of a dock. Since that time dock has followed dock, and the record of the port has been one of uninterrupted and phenomenal growth. Cardiff now possesses five docks, besides timber floats and graving-docks. The total area covered by the wet docks is 163 acres.

The imports consist of iron ore, pig iron, timber, pit wood, grain, wood-pulp, flour, potatoes, and general merchandise. In 1912 they amounted in the aggregate to 2,096,392 tons. The exports consist of coal, coke, patent fuel, steel rails, pig iron, bricks, and general merchandise. In 1912 they reached the large total of 10,405,579 tons. The figures for the Customs port (including Penarth and Barry) in 1909 place Cardiff first of all the ports in the kingdom as regards foreign tonnage. Its record for 1912 was 8,170,231 tons as against 6,928,920 tons cleared from London. In the export of coal Cardiff easily takes the premier position in the world, its record for 1912 being 20,084,021 tons compared with 16,232,462 tons exported from the Tyne ports. Of this huge total the local port of Cardiff contributed not quite half.

The Penarth docks comprise a dock and a basin, constructed respectively in 1865 and 1881; they occupy together an area of 26 acres. Additional facilities for business are afforded by the tidal harbour alongside the docks, which has a total frontage of 15,000 feet and an area of 55 acres. The exports, which are similar to those of Cardiff, amounted in 1912 to 4,190,786 tons, of which over four millions were coal and coke. The imports, however, were only 110,080 tons.

The rise of Barry has been remarkable. In a quarter of a century it has grown from a rural hamlet to one of the largest coal ports in the Channel. The quantity of coal and coke exported in 1912 actually exceeded that shipped at Cardiff, the figures being 9,732,606 tons as

Barry Docks

against 9,601,648. The number of vessels cleared in 1912 was 3140, with an aggregate tonnage of 4,358,653 tons. Barry is at present little more than a coal port, but it has, perhaps, almost a finer prospect than any other port in the county. It owes its existence to a demand for greater shipping facilities than Cardiff could at the moment supply. The docks are three in

number, and cover a total area of 114 acres. There
are three large graving-docks and some timber ponds
of 41 acres. Barry possesses one advantage of which
none of the rival Channel ports can boast, for the docks
can be entered at all states of the tide, and owing to the
absence of rocks and shoals they can be approached with
ease. The exports in 1912 were 10,371,775 tons.

Porthcawl possesses a small dock of seven acres, and
once did a brisk trade in coal, but the dock has now been
dismantled and practically abandoned.

The decline of Porthcawl has been more than coun-
terbalanced by the rise of its neighbour Port Talbot at the
mouth of the Avon, which is now an exceedingly busy
place. It has two large docks covering together 92 acres,
and a large graving-dock. The harbour has a fine entrance,
but owing to the sandy nature of the locality it has to be
kept free by constant dredging. Port Talbot thrives largely
on its exportation of coal, but it has likewise secured a considerable import trade.

Briton Ferry possesses a small dock of 14 acres, which
is kept busily employed, and though this to a large extent
provides a harbour for Neath, yet vessels of less than 400
tons can proceed two miles up the river to Neath itself,
where they lie alongside the quays.

Swansea, though now outstripped by its ancient rival
Cardiff, is yet a port of very considerable importance and
magnitude. In some ways it is the most interesting port
in Glamorganshire, for its trade is far more varied than
that of any of the others. Though it exports a very
large quantity of coal, coke, and patent fuel, yet it is by

no means exclusively a coal port. One very important feature of its trade is the importation of the various metallic ores required by the different smelting works in the neighbourhood. Amongst the exports tin plates figure largely. In 1912 the imports amounted to 1,021,583 tons, and the exports to 5,282,590 tons, of which 3,898,135 were coal and coke. The number of vessels cleared from the port in 1912 was 5638, with a total tonnage of

King's Dock, Swansea

3,028,383 tons. The port possesses four docks covering a total area of 133 acres. The oldest dock was constructed in 1852 by damming the river Tawe, which has now to find its way to the sea by a new cut. The latest dock, the King's Dock—a splendid sheet of water—was opened in 1909. It can accommodate the largest vessel afloat and is entered by a fine deep-sea lock 875 feet in length. There are large floats and graving-docks.

17. History of Glamorganshire.

There is no recorded history of Glamorgan before the coming of the Romans. Together with the present counties of Brecon and Monmouth it was at the commencement of the Roman occupation in the possession of the Silures, a people of mixed Iberic and Goidelic race, who were eventually reduced by Sextus Julius Frontinus. He is reputed to have achieved the pacification of the country by the construction of an elaborate system of military works. In any case his operations were completely successful, and by A.D. 78 he had obtained a secure hold over South Wales, and it only remained for his successor, Agricola, to complete the subjugation of the North.

We are again without knowledge of the life of the people under the rule of their Roman masters. The absence of history is some evidence of the success of the Roman administration. Probably Wales was regarded as a disturbed area which required watching. The Roman legions were kept massed on the frontiers to overawe the unruly tribesmen. The headquarters of the troops were at Caerleon, and subsidiary camps were established at Cardiff and Gelligaer. The fact that the natives retained their own language shows that national feeling was by no means extinct, though doubtless customs were to some extent modified by contact with the Romans.

After the departure of the Romans at the beginning of the fifth century the most important event for South

Wales was the Brythonic invasion from the north.
Until the fifth century the Brythons do not appear to
have advanced further south than the territory of the
Ordovices in Central Wales. They allowed the Goidels
to remain in undisturbed possession of the south. On the
withdrawal of the Romans, however, the Brythons overran
Ceredigion and gradually pushed their way between the
Tawe and the Towy, but it was not until the sixth and
seventh centuries that they appear to have overflowed
into Dyfed and Morganwg. In the end so complete
was their conquest that all trace of the earlier Goidel
supremacy disappeared and with it the language.

The Brythons, however, were not long left in peaceful
possession of their new conquests. In the sixth century
another foe—the Saxon—had appeared on the eastern
shores of Britain. By the middle of the seventh century
they had reached the frontiers of Wales and it was probably
the pressure of this fresh invasion that gave the Welsh
whatever solidarity as a nation they have since possessed.
Iberian, Goidel, and Brython all became "Cymry," or
fellow-countrymen. We once more lack reliable information as to what actually took place. The early history
of Glamorgan is especially obscure. It was however
from Morgan Mwynfawr in the eighth century that this
part of the realm subsequently acquired its permanent
appellation of Morganwg. The Saxons seem to have
made but little advance into South Wales, though the
native princes were no doubt constantly employed in
repelling their attacks. In 720, for instance, Ethelbald
of Mercia is said to have descended on Gwent and

Morganwg and to have pillaged the district, but in spite
of these raids, Offa's Dyke on the English side of the
Wye remained the permanent line of demarcation
between the Welsh and their invaders. In the ninth
century the incursions of the Saxons were varied by the
ravages of the Danes. In 877 the "black pagans" are
reputed to have laid waste long stretches of land on the
fertile shores of Morganwg, and in 896 to have come
again and devastated Gwent and Gwynllwg. The Norse
names of some of the localities on the coast probably
mark the temporary sojourning places of the sea-rovers.

The Scandinavian peril put an end for a time to
English and Welsh antagonism, and with a view to
protecting their territories from these Danish onslaughts
the South Wales princes put themselves under the protection of Alfred the Great, whose victory over the
common foe in 878 made them look to him as their
champion in the hour of distress. These good relations
were maintained with the English during the reign of
Edward the Elder, who gave proof of his friendliness to
the Welsh by materially helping them during a vigorous
Danish raid on the coast of Morganwg in 915. In spite
of these vicissitudes Morganwg remained under the rule
of its own princes, the descendants of Hywel ap Rhys,
till the eleventh century. In 1043 a Gwentian prince,
in the person of Gruffydd ap Rhydderch, came to the
throne and extended his power over the neighbouring
kingdom of Deheubarth. His dynasty, however, was
short lived. The last native prince was Iestyn ap Gwr-
gant, who seized the realm on the fall of Gruffydd's son

Caradoc. It was at the hands of Iestyn that Morganwg
lost its independence. Iestyn's downfall was occasioned
by the arrival of the Normans. The real history of the
Norman conquest of Glamorgan is strangely obscure.
It is difficult to unravel the facts from the fiction with
which they are interwoven. The story is told with
much circumstantiality by the Welsh chroniclers, but
their narratives are quite at variance with the little that
is really known.

Some sort of an agreement seems to have been arrived
at between William the Conqueror and Rhys ap Tewdwr,
the ruler of Deheubarth, by which the Welsh prince
secured undisturbed possession of his dominions in return
for a fixed tribute ; but the Conqueror's death put an
end to the arrangement. The hand of Rufus was not
strong enough to hold back any longer the turbulent
Norman barons, and Rhys fell in battle in a vain attempt
to prevent the Normans from entering Breconshire.
In 1093 Fitzhamon, a kinsman of Rufus, who already
possessed the lordship of Gloucester, pushed westwards ;
and in an incredibly short time the whole of the district
between the Usk and the Avon was in his hands. In his
wake came a swarm of Norman adventurers who carried
their standards across Gower into Carmarthenshire. By
1094 a chain of fortresses extended from Newport to
Pembroke, and five years later Henry Beaumont, Earl of
Warwick, added to the Norman dominions the peninsula
of Gower. Not only were the lowlands of Glamorgan
in the possession of Fitzhamon and his followers, but the
mountainous district of the north was at least nominally

subject to his authority. What the Norman could not conquer he was content to rule, and the Welsh were allowed to retain their hold upon Senghenydd—the district of Caerphilly—on the acknowledgment of Fitz-hamon's jurisdiction. The lordship of Glamorgan be-came thus practically coterminous with the old kingdom of Morganwg. Richard de Grenville secured an inde-pendent sphere on the banks of the Nedd, and the land between the Nedd and the Avon, perhaps on account of its mountainous character, was allotted to a Welsh chief-tain, Caradoc ap Iestyn.

The subsequent history of Glamorgan is a record of the efforts of the Normans to retain possession of the territory they had won. Fitzhamon died in 1107 and was succeeded by Robert of Caen, the natural son of Henry I, who had married Fitzhamon's daughter. He died in 1147, and was succeeded by his son Earl William. Henry II twice traversed Glamorgan with a view to advertising his sovereignty, but his visits did little to sober the restlessness of the Welsh. The death of Earl William in 1183 was followed by a local rising under the leadership of Caradoc, the Welsh lord of Avon, in the course of which Cardiff and Kenfig were burnt and the security of Neath threatened. The close connection which already existed between the lordship of Glamorgan and the English crown was for a while renewed by the succession of Prince John to the Welsh estates of Earl William, whose heiress he married ; but on John's contracting a second marriage the lordship passed to Richard de Clare, the husband of another daughter. For more than a century Glamorgan remained in the hands of this powerful house, and its history is bound up with the story of their fortunes.

In the thirteenth century occurred the elder Llewelyn's struggle to secure Welsh independence, but though the prince ravaged Gower in the interest of his son-in-law, John de Breose, and kept the country generally in a state of unrest, no event of any importance took place on Glamorgan soil. Under the leadership of the younger Llewelyn, the English possessions in Glamorgan were threatened. In 1257 the Welsh prince burst into the county and destroyed the castle of Llangynwyd, and in 1271 demolished the castle which Gilbert the Red had begun at Caerphilly. But though Llewelyn was allowed to retain for a while his hold upon the northern half of Meisgyn, the final hopes of Welsh liberty were extinguished when Edward I crushed his power and pushed his victorious arms to Carnarvon. The conquest of the rest of Wales in nowise altered the political status of Glamorgan. It still remained for all practical purposes the personal possession of the de Clares, though subject nominally to the supremacy of the crown. On the death of the last de Clare at Bannockburn the lordship again changed hands. By a matrimonial alliance the inheritance was secured by the Despensers. Once more there was trouble on the Welsh border. In 1315 the rapacious Hugh le Despenser annexed the district of Senghenydd, which had hitherto been regarded as a semi-independent Welsh lordship. Llewelyn Bren, the aggrieved chieftain, appealed for justice to the English

crown. Failing to obtain redress he collected a considerable army and laid waste the land of Glamorgan.
So threatening became the situation that the Earl of
Hereford was deputed to put down the rebellion. Llewelyn,
realising the hopeless nature of the struggle, retired to the
mountains and after a short resistance capitulated. He
was pardoned and liberated, but falling subsequently into
the hands of Sir W. Fleming, Despenser's lieutenant,
he was dragged through the streets of Cardiff, hanged
and quartered. For this act of lawlessness Fleming was
himself afterwards gibbeted—a circumstance which shows
that the Crown sometimes made its authority felt within
the limits of the lordship. The exactions of Despenser
soon raised against him a more formidable foe than
Llewelyn Bren. The Earl seized Caerphilly, which was
claimed by Roger Mortimer, and some further attempts
to enrich himself at the expense of other nobles led to the
formation of a powerful coalition of Marcher lords who
were bent upon his destruction. This private feud set
the whole kingdom in a blaze. Edward II's refusal to
get rid of his favourite caused the outbreak of hostilities.
The queen, who had made common cause with the dis-
contented nobles, collected an army and pursued the king
to Bristol, where the elder Despenser fell into her hands.
The king with the younger Despenser took to flight and
made an attempt to reach Lundy Island. He was, however, driven on to the coast of Glamorgan, and after
shutting himself up for a time in the castle of Caerphilly,
eventually sought an asylum in the abbey of Neath.
Again he made a futile effort to escape, but with

Caerphilly Castle

Despenser fell into the hands of his enemies at Llantrissant. Despenser was at once hanged and the lordship of Glamorgan was conferred on Mortimer.

The next political upheaval which convulsed the county was the insurrection of Owain Glyndwr. Owain had no claims to sovereignty, though he sprang from royal blood, but he was not without political ideals, and he possessed a contagious enthusiasm which won all hearts. His almost invariable good fortune and personal adroitness obtained for him amongst the imaginative Welshmen the reputation of a necromancer. The opening of his career was like that of Llewelyn Bren. He flew to arms by way of reprisal for some personal injury, and the success which attended his earlier movements kindled in him the ambition to become a national deliverer. At one time he seemed likely to achieve his desire. A campaign of plunder in South Wales and a victory over Lord Grey at Pilleth in 1402 awoke once more the dream of Welsh independence, and Glyndwr became the idol of his countrymen. He followed up his victory by ravaging Gwent and Glamorgan. He burnt the town of Cardiff and destroyed the Bishop's palace at Llandaff. The next year the revolt of the Percys, who had allied themselves with their old antagonist, seemed to put success almost within his grasp; but the defeat of the younger Percy at Shrewsbury dashed his hopes. The disaster did not, however, cause Glyndwr to relax his hold upon South Wales. All the castles from Newport to Pembroke were in distress, and the King made a futile expedition as far as Carmarthen to relieve them. In

1404, according to local testimony, Glyndwr was again
in Glamorgan, and after destroying the castles of Penlline,
Llandough, Llanblethian, Flemingston, Talyfan, Malafant, and Penmark, he gained a sanguinary but indecisive
victory over his opponents at Stalling Down near Cowbridge. This proved his last substantial success. In 1405
the Welsh chieftain's star began to wane. Prince Henry
defeated his army in both Gwent and Brycheiniog. His
exploits had been brilliant rather than fruitful, and these
reverses, coupled with his previous failure to make any
real headway, destroyed his prestige. The politic offer of
liberal terms to deserters caused his army gradually to
melt away, and though for some years afterwards he
continued to give trouble, his cause was hopelessly ruined.
After carrying on a desultory and fitful guerrilla warfare
amongst the mountains until 1413, he eventually came
to a nameless grave.

On the accession of Henry VII the Welsh in a sense
came to their own again. Henry was on his father's
side a Welshman, and one of the first things he did was
to confer the lordship of Glamorgan, of which Richard III
had been the last holder, upon his uncle Jasper Tudor.
Jasper's rule proved popular, and did something to prepare
the way for the final incorporation of Wales with the
English crown. In the reign of Henry VIII a change
in administration was effected. The county of Glamorgan formally became a shire, and its boundaries
were readjusted. The district of Gwynllwg was given
to Monmouthshire and the peninsula of Gower was
added to Glamorganshire. The gain to the county was

considerable. The shire became entitled to send representatives to the English Parliament, and all lawsuits came under the direct cognisance of the English courts.

In 1539 the Glamorganshire monasteries, which had already experienced the heavy hand of Glyndwr, were suppressed, and with them fell the friaries, which, on account of their popular sympathies, Glyndwr had spared. In the days of Elizabeth, Cardiff got into bad repute. Owing to the expansion of local commerce the Bristol Channel became infested with pirates, who preyed on the sea-going vessels. Their depredations were connived at by the people of Cardiff, who found in the disorganisation of trade an excellent opportunity for cheating the revenue. The town became notorious as the resort of desperados, and the town authorities were not without reason suspected of encouraging their illicit practices. Sir T. Button was commissioned to put an end to this illegal traffic, but the smuggling went on to the end of the century, when Edmund Mathews was prosecuted for clandestinely supplying the Spaniards with cannon.

In the Stuart period the county again became involved in political turmoil. As a whole it was strongly royalist, and it loyally responded to the king's demand for ship-money, probably under the mistaken impression that the ships were to be used for the protection of the maritime trade. The enthusiasm, however, gradually gave place to indifference, which soon developed into sullen opposition. On the outbreak of hostilities the popular resentment against these frequent imposts began to display itself, and an additional cause of offence was given when, after a disastrous skirmish near Tewkesbury, in which the Glamorganshire levies had suffered severely, the Marquis of Hertford was replaced as commander of the South Wales forces by Lord Herbert, a papist. The royal cause was rendered still more unpopular by the exactions of Colonel Gerard, who had been despatched by Prince Rupert to clear Glamorganshire of the Parliamentary forces. After the decisive battle of Naseby the King became a fugitive and sought a temporary asylum at Raglan. From thence he journeyed to Cardiff in the hope of raising a fresh force in Glamorgan. But Charles was soon made aware how completely he had alienated public sympathy. A formal meeting was arranged between the King and the people at St Fagan's, but the demeanour of the crowd was so threatening that Charles withdrew in fear from the conference. After some further negotiations in which the King undertook to grant all the popular demands, a half-hearted promise of support was given. But it was not kept, and the royal fugitive had to leave without the army which he came to seek. After the final collapse of the King's cause, a futile attempt was made by the royalists in 1648 to take advantage of the dissensions between the Independents and Presbyterians and to effect a fresh rising. Siege was laid to Cardiff Castle, but the investing force was easily dispersed at the approach of the Parliamentary army. In the following year, however, a sudden revival of royalist hopes was brought about in an unexpected way. Major-General Laugharne, the commander of the Parliamentary forces in South Wales, disappointed, it is said, at the inadequate reward of his services, forsook his party.   He carried over with him not only his officers but his army. Cromwell himself hastened to Wales to crush the dis- affection.   The rebellion was over, however, before he arrived.   Laugharne advanced from Pembroke into Gla- morgan, and encountering Colonel Horton, who had been sent forward by Fairfax to intercept him, at St Fagan's, at once attacked him.   The combat was most sanguinary, and ended in the complete defeat of Laugharne and the practical annihilation of his army.   The battle of St Fagan's was the last event of national importance in the annals of Glamorgan.   The subsequent story of the shire is the history of its rapid commercial development, and of its religious revival through the efforts of the early Methodists.   The only incidents which have since oc- curred to break the peace of the county were industrial riots in the Merthyr district in 1802 and 1816 consequent upon the lowness of wages.

18. Antiquities.

We have no written record of the history of our land antecedent to the Roman invasion in B.C. 55, but we know that Man inhabited it for ages before this date. The art of writing being then unknown, the people of those days could leave us no account of their lives and occupations, and hence we term these times the Prehistoric period.   But other things besides books can tell a story, and there has survived from their time a vast quantity of objects (which are daily being revealed by the plough of the farmer or the spade of the antiquary), such as the weapons and domestic implements they used, the huts and tombs and monuments they built, and the bones of the animals they lived on, which enable us to get a fairly accurate idea of the life of those days.

So infinitely remote are the times in which the earliest forerunners of our race flourished, that scientists have not ventured to date either their advent or how long each division in which they have arranged them lasted. It must therefore be understood that these divisions or Ages—of which we are now going to speak—have been adopted for convenience sake rather than with any aim at accuracy.

The periods have been named from the material of which the weapons and implements were at that time fashioned—the Palaeolithic or Old Stone Age; the Neolithic or Later Stone Age; the Bronze Age; and the Iron Age. But just as we find stone axes in use at the present day among savage tribes in remote islands, so it must be remembered that weapons of one material were often in use in the next Age, or possibly even in a later one ; that the Ages, in short, overlapped.

Let us now examine these periods more closely. First, the Palaeolithic or Old Stone Age. Man was now in his most primitive condition. He probably did not till the land or cultivate any kind of plant or keep any domestic animals. He lived on wild plants and roots and such wild animals as he could kill, the reindeer being then abundant in this country. He was largely a cave-dweller and probably used skins exclusively for clothing. He erected no monuments to his dead and built no huts. He could, however, shape flint implements with very great dexterity, though he had as yet not learnt either to grind or to polish them. There is still some difference of opinion among authorities, but most agree that, though this may not have been the case in other countries, there was in our own land a vast gap of time between the people of this and the succeeding period. Palaeolithic man, who inhabited either scantily or not at all the parts north of England and made his chief home in the more southern districts, disappeared altogether from the country, which was later re-peopled by Neolithic man.

Neolithic man was in every way in a much more advanced state of civilisation than his precursor. He tilled the land, bred stock, wove garments, built huts, made rude pottery, and erected remarkable monuments. He had, nevertheless, not yet discovered the use of the metals, and his implements and weapons were still made of stone or bone, though the former were often beautifully shaped and polished.

Between the Later Stone Age and the Bronze Age there was no gap, the one merging imperceptibly into the other. The discovery of the method of smelting the ores of copper and tin, and of mixing them, was doubtless a slow affair, and the bronze weapons must have been ages in supplanting those of stone, for lack of intercommunication at that time presented enormous difficulties to the spread of knowledge. Bronze Age man, in addition to fashioning beautiful weapons and implements, made good pottery, and buried his dead in circular barrows.

In due course of time man learnt how to smelt the ores of iron, and the Age of Bronze passed slowly into the Iron Age, which brings us into the period of written history, for the Romans found the inhabitants of Britain using implements of iron.

We may now pause for a moment to consider who these people were who inhabited our land in these far-off ages. Of Palaeolithic man we can say nothing. His successors, the people of the Later Stone Age, are believed to have been largely of Iberian stock; people, that is, from south-western Europe, who brought with them their knowledge of such primitive arts and crafts as were then discovered. How long they remained in undisturbed possession of our land we do not know, but they were later conquered or driven westward by a very different race of Celtic origin—the Goidels or Gaels, a tall, light-haired people, workers in bronze, whose descendants and language are to be found to-day in many parts of Scotland, Ireland, and the Isle of Man. Another Celtic people poured into the country about the fourth century B.C.—the Brythons or Britons, who in turn dispossessed the Gael, at all events so far as England and Wales are concerned. The Brythons were the first users of iron in our country.

The Romans, who first reached our shores in B.C. 55, held the land till about A.D. 410; but in spite of the length of their domination they do not seem to have left much mark on the people. After their departure, treading close on their heels, came the Saxons, Jutes, and Angles. But with these and with the incursions of the Danes and

Irish we have left the uncertain region of the Prehistoric Age for the surer ground of History.

The earliest stone implements in Wales were found on the floors of the caves of Gower. Beneath layers of stalagmite were discovered a number of flint flakes and a fine flint arrow-head. The rough unpolished nature of the weapon has caused it to be assigned to the Palaeo-

Neolithic Implements found at Cowbridge

lithic or early Stone Age, when men were accustomed to chip instead of polishing their tools. Palaeolithic remains are, however, extremely scanty in South Wales. Of the Neolithic or later Stone Age relics are more numerous. A fine polished axe-head, now in the British Museum, was found at Cardiff, and another good specimen pierced with a hole for a handle was turned up at Llanmadoc in Gower. It is now commonly supposed that the men who

used the polished stone tools were of the Iberian race.
The Goidels who succeeded them were furnished with
implements of bronze. Their weapons were the axe,
sword, spear, and knife. The specimens found vary
considerably in efficiency, and show that the early Celtic
artificer only gradually learnt the mastery of his craft.
It was left to the Brython to bring with him the more
effective iron weapon. The Brython had been eight or
nine centuries in Britain before he penetrated into South
Wales, but there is no reason to suppose that the Goidels
of Glamorgan retained the use of bronze weapons until
the Brythonic conquest in the fifth century of our
era. They must have learnt to some extent the use of
iron from contact with their Brythonic neighbours in the
centre of Wales. Numerous specimens of iron and bronze
instruments are to be seen in the Cardiff Museum.

Besides the implements and fragments of pottery discovered casually in different localities, there are several
prehistoric monuments in various parts of the county.
Glamorganshire as a whole is particularly rich in such
remains. The chief objects of antiquarian interest are
encampments, cromlechs, barrows, cairns, stone circles,
and menhirs or long stones.

Encampments are very numerous, and are to be found
on the summits of many of the hills, and in one or two
elevated positions on the shore. They consist of a series
of embankments and ditches thrown round the top of a
hill or drawn across the corner of some projecting cliff to
render the situation more difficult of approach. Probably
in time of warfare the tribesmen gathered their cattle and

other belongings within these enclosures and further fortified them by the erection of a wooden stockade. It is
difficult to assign these earthworks to any particular period.
Probably they were used in all periods. The positions

St Lythan's Cromlech

were no doubt re-fortified more effectually as each new
race succeeded to the inheritance of the old. The form
of the camp varied with the nature of the ground, though
the Romans seem invariably to have made their entrench-
ments rectangular.

Next to these early methods of warfare we know most about ancient methods of sepulture. The principal sepulchral monuments are cromlechs (or dolmens), barrows, cairns, and menhirs. A *cromlech* is a group of stones arranged like a table. A flat slab is supported horizontally on three or more upright stones. Together, these formed a chamber in, or under which the dead man was placed, the corpse being sometimes supplied with his weapons and a food vessel. The cromlech was in some cases covered with a mound of earth. In the case of most cromlechs, however, the earth has either been removed or never existed, and the chamber is exposed. The largest cromlech in Great Britain is near St Nicholas, and this still retains evident traces of the mound which once covered it. In the same neighbourhood, near Dyffryn House, is another very fine specimen, which is completely exposed; and a third well-known example is Arthur's Stone on the summit of Cefn-y-Bryn in Gower. It is evident from the unwrought condition of the stones that they belong to a very early age, and they are now generally assigned by antiquaries to the Neolithic or Iberian period.

The Celts introduced cremation. The ashes were enclosed in a vessel and then placed in a small chamber built like a miniature cromlech and called a *cist-faen*. Some of these little stone chests were surrounded by a circle of upright stones. Though the large stone circles like Stonehenge and Avebury may have been used for worship and were possibly also rude astronomical appliances, there is little doubt that the smaller circles, such as those found in Glamorganshire, were merely sepulchral

monuments.   They probably belong to the Bronze or
Goidelic period.   The chief stone circles in the county
are on Carn Llechart near Llangyfelach, on Cefn-y-
Gwrhyd near Llangiwg, and on Drummau Hill near
Neath.

Sometimes the bodies or ashes of the dead were placed

King Arthur's Stone

in mounds or *barrows*.   A great number of these exist in
different parts of the county, generally on the hill tops,
or in the vicinity of some ancient trackway.   When the
barrow contains a chamber it is probably an Iberic tomb;
if it contains no chamber, it is probably of Celtic construction.   In some cases where the earth was scanty,

the grave was covered with a heap of stones called a *cairn*, though many of the cairns seen on the tops of the highest hills are merely landmarks. A barrow of very large size is generally called a *tumulus*, and was probably a place of general sepulture, though some of them may have been ancient signalling stations. A very fine tumulus occurs on the top of Crug-yr-Afon in the centre of the county.

The *menhir* (*maen-hir* or long stone) is another form of sepulchral monument, also believed to be of Goidelic

origin. It is simply an unhewn or roughly-shaped stone or slab placed upright in the earth, often of large size. Of these there are many in the county. Very interesting are certain stones with a roughly-cut inscription giving the dead man's name in Welsh or Latin or sometimes in both languages. The most remarkable examples

Ogam Stone near Kenfig

of the kind in Glamorganshire are the stones at Kenfig and Loughor. These have the epitaph cut not only on the face of the stone in Latin, but in "Ogams" as they are called, notches cut on either side of the edge of the stone. These characters are now recognised as a Gaelic method of writing. As they were executed in post-Roman times, it is evident that the Goidelic language could not have disappeared from Wales until after the departure of the

Romans. It is argued from this fact that the Brythonic tribes did not overrun South Wales until the fifth century

The Goblin Stone

A.D. Some, however, maintain that the stones may have been erected by immigrants from Ireland.

Amongst the more notable of the other inscribed stones scattered about the county are the Bodvoc stone on Margam mountain, the Tome stone at Margam, the Teyrnoc stone on Cefn Brithdir, a defaced stone on Cefn Gelligaer, and the Macaritinus stone near Neath.

The relics of the Roman period are few. They consist chiefly of encampments and the foundations of villas and military fortifications. Three Roman roads traversed Glamorgan, the *Via Julia* which came out of Monmouthshire and skirting the coast crossed the lowlands from Cardiff to Loughor; the *Sarn Hir* which ran over Gelligaer mountain from Cardiff to Brecon; and the *Sarn Helen*, which traversed the vale of Neath and connected Neath with Brecon. It is very doubtful whether these roads were originally of Roman construction. They were probably ancient trackways utilised by the Romans for the movements of troops. There were four military stations along the route of the *Via Julia*— Tibia Amnis, Bovium, Nidum, and Leucarum. Tibia Amnis is now generally identified with Cardiff. That Nidum and Leucarum were Neath and Loughor is suggested by the similarity of the names, but there is no agreement about the situation of Bovium. Cowbridge, Bonvilston, and Boverton all claim to have been the site.

The discovery of the foundations of a gateway, some bastions, and portions of a Roman wall surrounding the grounds of Cardiff Castle prove, as former rectangular embankments suggested, that the castle stands on the site of a considerable Roman fortification. More recent

Roman baths, Gelligaer

excavations have disclosed another Roman fortress and
the fragments of some baths near Gelligaer. Both places
were probably permanent garrison stations on the Welsh
frontier. Roman camps exist at Caerau and Bonvilston,
on the supposed line of the *Via Julia*; also at Penydarren
and near Colbren on the north border, on Mynydd-
y-Gaer near Bridgend and on Mynydd Margam. The
foundations of Roman buildings have been unearthed
at Ely near Cardiff, and at Llantwit Major, but the
infrequency of such discoveries, and the general absence
of evidences of domestic life, tend to show that the
county was occupied by the Romans only in a military
sense, and that the ancient inhabitants were but little
influenced by Roman civilisation, though the Latin inscriptions on the standing stones prove that they must
have acquired some Roman customs.

19. Architecture—(*a*) Ecclesiastical.

Though Glamorganshire possesses one or two notable
churches, its display of ecclesiastical architecture is, on the
whole, disappointing. The reason for this is not far to
seek. Architecture depends for its development both
upon an adequate supply of building material and upon
the existence of a body of skilled workmen. There is
an abundance of stone in Glamorgan, but it does not
lend itself to the finest manipulation; and the political
condition of the county was too disturbed to give the
Welsh much chance of acquiring the art of building.
The English practically learnt their architecture from

the Normans, who were the most accomplished craftsmen of the day. But while in England the Norman conquest eventually took the form of a racial amalgamation, in Wales it remained a foreign occupation. The Welsh had little social and religious intercourse with their Norman masters. The buildings which the Normans themselves reared were either castles, or monasteries for the accommodation of foreign monks. Ecclesiastical architecture as a rule owed its growth to the influence of the monasteries. They supplied the resources and trained the workmen. But the Welsh disliked the monks almost as cordially as they hated the barons. They regarded them as foreign garrisons in spiritual guise; so that Welsh architecture profited but little by the existence of the monasteries.

The Norman conquest, by the dislocation of national life, likewise prevented Welsh ecclesiastical art from developing on its own lines. That it had a tradition of its own is abundantly proved by the existence of the large number of elaborately-worked crosses which are scattered up and down the Principality. Glamorganshire is rich in them. There are two especially fine collections of them within the county, one at Llantwit Major and the other at Margam. Both of these places were active centres of church life before the coming of the Norman, and both must have possessed schools of skilled workmen. The crosses usually take the form of an upright pillar or a wheel cross, highly ornamented with an intricate pattern of plait work in bold relief, and, as their inscriptions show, they were mostly sepulchral monuments. Sometimes they bear in addition a rude attempt at figure sculpture. They are probably not earlier than the eighth or ninth centuries, and many of them are mutilated. Besides the groups at Margam and Llantwit there are some good specimens at Merthyr Mawr, Coychurch, and Llangan. At the latter place the cross exhibits a crude representation of the Crucifixion. In the churchyard of Llandough near Cardiff there is a fine but imperfect specimen of a figured cross. No Celtic churches have come down to us, but many of the existing churches were no doubt built on sites consecrated by ancient ecclesiastical associations. At Llantwit Major and Llancarfan were Welsh monastic institutions, and the cathedral of Llandaff perpetuates the Welsh episcopal traditions of the place. There are some fine examples of Norman and English ecclesiastical architecture within the county, but the best buildings are, as we should naturally expect, monastic in origin. The parish churches, as a rule, are small and poor.

A preliminary word on the various styles of English architecture is necessary before we consider the churches and other important buildings of our county.

Pre-Norman or (as it is usually, though with no great certainty termed) Saxon building in England was the work of early craftsmen with an imperfect knowledge of stone construction, who commonly used rough rubble walls, no buttresses, small semi-circular or triangular arches, and square towers with what is termed "long-and-short work" at the quoins or corners. It survives almost solely in portions of small churches.

The Norman conquest started a widespread building of massive churches and castles in the continental style called Romanesque, which in England has got the name of "Norman." They had walls of great thickness, semi-circular vaults, round-headed doors and windows, and massive square towers.

From 1150 to 1200 the building became lighter, the arches pointed, and there was perfected the science of vaulting, by which the weight is brought upon piers and buttresses. This method of building, the "Gothic," originated from the endeavour to cover the widest and loftiest areas with the greatest economy of stone. The first English Gothic, called "Early English," from about 1180 to 1250, is characterised by slender piers (commonly of marble), lofty pointed vaults, and long, narrow, lancet-headed windows. After 1250 the windows became broader, divided up, and ornamented by patterns of tracery, while in the vault the ribs were multiplied. The greatest elegance of English Gothic was reached from 1260 to 1290, at which date English sculpture was at its highest, and art in painting, coloured glass making, and general craftsmanship at its zenith.

After 1300 the structure of stone buildings began to be overlaid with ornament, the window tracery and vault ribs were of intricate patterns, the pinnacles and spires loaded with crocket and ornament. This later style is known as "Decorated," and came to an end with the Black Death, which stopped all building for a time.

With the changed conditions of life the type of building changed. With curious uniformity and quickness the style called "Perpendicular"—which is unknown abroad—developed after 1360 in all parts of England and lasted with scarcely any change up to 1520. As its name implies, it is characterised by the perpendicular arrangement of the tracery and panels on walls and in windows, and it is also distinguished by the flattened arches and the square arrangement of the mouldings over them, by the elaborate vault-traceries (especially fan-vaulting), and by the use of flat roofs and towers without spires.

The medieval styles in England ended with the dissolution of the monasteries (1530—1540), for the Reformation checked the building of churches. There succeeded the building of manor-houses, in which the style called "Tudor" arose—distinguished by flat-headed windows, level ceilings, and panelled rooms. The ornaments of classic style were introduced under the influences of Renaissance sculpture and distinguish the "Jacobean" style, so called after James I. About this time the professional architect arose. Hitherto, building had been entirely in the hands of the builder and the craftsman.

Llandaff Cathedral is not only the principal but also the most noteworthy ecclesiastical building in the county. It is one of the few cathedrals built strictly in the form of a parallelogram. It is a long structure of many styles standing on the right bank of the Taff. It was commenced by Bishop Urban in 1120 on the site of some earlier fabric and underwent constant alteration and extension up to the end of the fifteenth century. After the Reformation, when it was despoiled of its possessions, it fell into decay, from which it was not rescued until

Llandaff Cathedral

the middle of the last century. It consists of a nave, choir, chapter-house, and Lady chapel. Its finest feature is the west front, which is flanked by two lofty towers. The earliest portions of the church are the richly sculptured Norman archway at the east end of the choir and the two Norman doorways in the north and south walls of the nave. The nave and west front are Early English, and the chapter-house, which has a characteristic vault, is of the same period, but rather later in date. The Lady chapel, which is also vaulted, is a good example of early Decorated work, and was built between 1285 and 1287, when the earlier Gothic architects were feeling their way to a more elaborate style. The presbytery and aisles were remodelled a little later, when the Decorated style was fully developed. The only feature belonging to the Perpendicular period is the north-west tower, which is said to have been erected by Jasper Tudor. Its companion tower, which carries a spire, is modern.

Next to Llandaff Cathedral, the finest church in the county is Ewenny Priory. It is the most splendid example of pure Norman architecture in Wales. The building is cruciform, with a massive central tower, and consists of nave, choir, and south transept. The north transept has been demolished, but otherwise the fabric is unimpaired. The barrel-vaulting of the choir is particularly noteworthy. With the exception of the gateway the monastery has been demolished.

Margam church is another example of Norman work. It was once the nave of a large Cistercian monastery, the choir of which has been destroyed. Adjoining it, however, are the ruins of a beautiful chapter-house, which is a splendid specimen of Early English architecture.   The foundations of the dismantled choir and some fragments of the domestic buildings still remain.

Neath Abbey, which is traditionally reputed to have been the work of the same architect, is of Early English

Ewenny Priory Church

style.   The church, which was once regarded as one of the most splendid ecclesiastical edifices in Wales, is very ruinous, but a large vaulted chamber in the adjacent buildings is in a good state of preservation.   A most interesting but rather uncouth structure is the church of Llantwit Major.   It belongs chiefly to the thirteenth and fifteenth centuries, and exhibits the usual monastic arrangement. The building is divided into two by a central tower. At the west end are the ruins of a Lady chapel, with an adjoining chamber, and on the rising ground above are a gateway and a circular columbarium. A priory and some habitations of Black and Grey Friars existed at Cardiff, but these have disappeared, though the foundations of the latter can still be traced.

St Illtyd's, Llantwit Major

Of the parochial churches the most notable are Llancarfan, Coity, and Coychurch. There are Norman doorways at Marcross and Rhossili; Llancarfan has some Transitional Norman arcades; Cheriton is a good example of Early English; Coity, Coychurch, and St Fagan's are striking examples of the Decorated period; and St John's, Cardiff, is Perpendicular and has a particularly fine tower.

A peculiar feature of some of the Glamorganshire churches is their provision for defence in case of necessity. Ewenny Priory was enclosed by fortifications, and the church itself was made to minister to the security of the monks. The nave is shut off from the rest of the buildings by a stone wall, and the battlements of the massive

Newton Church

tower are pierced with loopholes for arrows. The tower of Newton Nottage is built almost like a castle keep and still shows the brackets for the support of the wooden staging which accommodated the defenders. The churches on the Gower coast have likewise towers capable of affording security in times of peril.

Churchyard Cross, St Donat's

Sketch Map showing the Chief Castles of Wales and the Border Counties

Woodwork is rare in the county. There is a poor screen and some stalls at Cowbridge, and the dilapidated remnants of a once rich screen at Llancarfan. Gileston church has a fine old door. Llantwit Major possesses a beautiful stone screen, and there is a curiously carved stone pulpit at Newton Nottage. Llanmaes church shows some medieval frescoes, and there are some later mural paintings at St Donat's. Several churches retain portions of their churchyard crosses, but only Llangan, St Donat's, and St Mary Hill preserve them in their entirety.

20.   Architecture—(*b*) **Military.**

The military architecture of Glamorganshire offers a very striking contrast to the ecclesiastical. The churches are few and poor, the castles were numerous and splendid. There were over 40 within the confines of the county. They were not, however, all residential castles ; some were merely garrison stations or military block-houses. They were all the work of the Norman conquerors or their descendants, though the ancient Welsh princes are credited with having had castles of some sort. The Normans had their own military methods, and the first thing they did to secure the foothold they obtained in Welsh territory was to erect at frequent intervals and in suitable positions temporary fortresses of turf and stakes. These were moated mounds crowned by a stockade and enclosed by a rampart. Upon the mound was eventually

erected a stone donjon or keep.   The distribution as well
as the design of these fortresses was determined by
strategical considerations.   They were part of a system
which had in view the defence of the whole territory as
well as the protection of personal estates.   The Normans
had not been long in the country before they drew a
string of permanent castles right across the southern

Fonmon Castle

portion of the county from the Taff to the Loughor
river.   Fortresses were also erected to guard the passes
from the hills and to watch the approaches from the
shore.

The typical Norman fortress consisted of a strongly-
built rectangular keep, which contained the living rooms
and store cellars.   This was generally surrounded by an
outer court protected by high and embattled walls and

encircled by a moat. The construction naturally varied with the peculiarities of the site, which was chosen both for its importance and inaccessibility. Various improvements in the art of castle-building were made from time to time as experience suggested. In the times of Richard I and John the keep became circular in form, and drum towers were placed at the angles of the outer court, which enabled the garrison to clear the walls of

Oystermouth Castle

assailants. In the reign of the Edwards a new type of fortress was introduced. The keep was first removed from the centre and placed upon the walls, and the enclosure was divided into two courts; but towards the end of the period the keep was altogether discarded, and the courts instead of being adjacent were made concentric, the dwelling rooms being ranged round the inner ward. The fortifications were buttressed with drum towers and the gateways were much strengthened. Many of the later castles were elaborate and extensive structures in which comfort as well as security was considered, and were provided with suites of apartments for the owner's family, a large banqueting hall, and a chapel. The original Norman castles often underwent reconstruction or were rebuilt in later times.

Coity Castle and Church

Most of the Glamorganshire castles are now in ruins. Some, like St George's and Scurlage, have quite disappeared ; others, like Sully and Llangynwyd, have been razed to the foundations. The hill forts of Castell Morgraig and, perhaps, Morlais seem never to have been completed. Cardiff and Castell Coch have been recently reconstructed, and Fonmon and St Donat's, though retaining some of their original features, have been transformed into residences and have seldom lacked an occupant. Some are noteworthy merely for their picturesqueness, whilst others are typical of their class. Ogmore preserves a fragment of its rectangular keep ; at Cardiff the keep is polygonal, but characteristic of its period. Oystermouth has some Decorated tracery. Coity possesses two adjacent courts, with the keep on the walls ; and Caerphilly (see p. 112) is a famous and typical example of a concentric castle of the most elaborate kind. Weobley is probably a late fortress. Some of the medieval towns were walled for defence. Cardiff, Kenfig, and Cowbridge were so protected. Kenfig has disappeared, and the fortifications of Cardiff have been destroyed, but Cowbridge still retains one of its gateways and a few fragments of walling.

21. Architecture—(c) Domestic.

We shall look in vain for early Welsh houses. A Welshman lived for the most part an out-of-doors life ; his ideas of personal comfort were limited and he went in constant fear of disturbance. His property therefore was always of a movable kind which could be packed up at the first alarm and carried off to the mountains. His house was merely a rude timber structure in which the chief and his men ate and slept together round the same hearth, with their horses near at hand. The Englishman's house, on the other hand, was literally his castle. It was both dwelling-place and fortress, for his security depended

not on the rapidity of his movements but on the strength
of his walls. Before the fifteenth century domestic architecture in Wales can scarcely be said to have existed.
The Welsh kept to their primitive dwelling in the mountains, and the Norman lord shut himself up in his castle.
The peasants herded in hovels round the strongholds, and
the houses of the townsmen were mostly of wood, and
were promptly pulled down in cases of fire. It was not
until the rise of the country gentry as a separate class
that domestic architecture underwent any great development.

The manor house as first constructed was simply a
large hall, the upper half of which was reserved for the
master and his family, and the lower allotted to the
servants. There were separate sleeping apartments, and
the larger halls were provided with a "solar" or with-
drawing room for the ladies, and a gallery for minstrels.
Though not military in character, the building was made
capable of defence and was sometimes moated. In Tudor
times, when the protection of the regular law did away
with the necessity of fortifications, personal comfort was
much more studied. Houses became more elaborate and
commodious. The household no longer lived a common
life, and privacy was obtained by adding to the number
of separate apartments. In the seventeenth century
architecture became more artificial. Architects who had
studied abroad came back with foreign ideas and began to
decorate houses with classical ornaments. This taste,
however, displayed itself chiefly in the more pretentious
buildings. Our domestic architecture never quite lost

touch with the Gothic traditions of the Middle Ages, and the smaller mansions and country cottages remained

Old Town Hall, Llantwit Major

picturesque and homely. In the reign of Queen Anne houses became more regular and stiffer in style, though they managed to preserve an air of stateliness. Later, architectural taste declined, and grace of outline was sacrificed to considerations of utility.

There are few early houses in Glamorganshire, though a number of Tudor mansions display earlier features. An example of an early residence which approximated to a castle is the ancient episcopal palace at Llandaff, which was destroyed by Glyndwr in 1402. Llantwit Major

Sker House

town-hall is a purely domestic building dating perhaps in substance from the fifteenth century. The quaint-looking hospice of St John at Bridgend is a fifteenth century building of the humbler kind. Llanmihangel Place was originally a fifteenth century castellated mansion, but Tudor gables have taken the place of the castellations and the great hall has been divided into separate

chambers.   Castell-y-Mynach, near Creigiau, though
chiefly a seventeenth century building, still retains a
chamber and an open timbered roof of the fifteenth
century.   Traces of fifteenth century work appear likewise in the walls of Flemingston Court, which was
once a fortified manor house, though its chief feature is a
typical Jacobean hall.   Characteristic examples of sixteenth century architecture are to be found in Nottage

Tudor Gardens, St Donat's Castle

Court, which contains some good panelled rooms,
Llancaich House, Sker House, and St Fagan's Castle,
a many-gabled mansion erected in 1578 within the
fortifications of an ancient castle.   Beaupré, which was
transformed from a castle into a manor house, exhibits a
beautiful renaissance doorway added to its inner court in
1600 by a Bridgend mason named Twrch.   St Donat's

Castle, a very composite building of almost all periods
from Norman times downwards, contains a good deal of
seventeenth century work, and some of its state apartments are embellished with the carving of Grinling
Gibbons.   Llansannor Court is a picturesque Jacobean
manor house.   Within the precincts of Neath Abbey are
the ruins of a mansion built in 1650 out of the wreckage
of the monastery, and incorporating some ecclesiastical
features.   The remains of the Herbert House at Cardiff,
built out of the material of the Grey Friars habitation,
comes down from Tudor times, as do the ruins of Llantrithyd Place.   A good example of a house illustrating
the architectural changes introduced in the eighteenth
century, is Cefn Mably on the banks of the Rhymney,
where the original gables have been replaced by dormer
windows.   Nash Manor is another Tudor building with
eighteenth century additions.   Ruperra Castle, though
popularly attributed to Inigo Jones, was rebuilt after a
fire in 1783, though it preserves its original seventeenth
century porch.   The Bishop's house at Llandaff, Briton
Ferry Manor, and the Gnoll near Neath, belong entirely
to the eighteenth century, and are very characteristic of
the time.

In the rural districts of Glamorgan the cottages have
a distinct character and conform more or less to one type.
They are low beetled-browed buildings, with thick squat
chimneys and a steep overhanging thatch, which curves
itself over the upper storey windows.   Their appearance,
though a little singular, is not unpicturesque.   An almost
universal feature of these humbler country homes is the

Neath Abbey

whitewash on their walls. The custom is of immemorial antiquity, and frequent reference to the "white cots" of Morganwg is made in the songs of the bards.

In the Gower peninsula a different type of cottage is seen. The dwelling is roofed with faggots, and thatched with sedge, and in the walls are bed-places.

22. Communications—Past and Present.

The art of road-making is in the British Isles a modern accomplishment, and Wales was one of the last districts where it made progress. In the prehistoric ages communication was established by trackways across the mountains, for a hill path afforded both directness and security, and loads could be carried on the backs of animals. Many of these ancient pathways, marked at intervals by burial mounds, can still be traced amongst the hills between the Rhondda and Avon rivers. The Romans were great road builders, though the roads which they built in England were comparatively few and were constructed chiefly for military purposes. It is doubtful whether they made any roads in Wales at all. Though three Roman roads have been mentioned, and were unquestionably used as highways, yet there is nothing to indicate that they were planned or paved in the usual Roman fashion. They probably ran along the line of some earlier trackway. Several portions of causeway, probably of Roman date, are, however, existent in the county.

From the time of the Romans almost to our own days Glamorganshire had practically no proper roads, and as late as 1780 the only coal brought to Cardiff was on horseback, and occasionally in barrows.   At the beginning of the nineteenth century a Parliamentary enquiry was held as to the character of the roads in South Wales, and a few main turnpike roads were constructed, in the face of

Cutting on the road near the Mumbles

much popular opposition, between the more important towns, but little was done to open up communication with the outlying districts of the north, except by the construction of canals and tramways.   The roads remained few and poor until the abolition of the turnpikes, and the establishment of a general system of road supervision.   Great improvements have been effected in the character of the roads within the past 30 years, but though much care is now bestowed upon their upkeep, Glamorganshire can perhaps hardly claim to be pre-eminent for its highways. As a rule they are somewhat narrow, though their surface is excellent. In the neighbourhood of the towns their improvement in condition is very marked. Two main highways run across the centre of the county from east to west; one puts Cardiff in connection with Neath and the western parts of the shire, the other with Bridgend by way of Llantrissant. Some good roads run up the valleys of the Taff and Nedd and put both Cardiff and Neath in communication with Merthyr, whilst another excellent high road leaves Swansea for Carmarthen. There are a large number of fairly good secondary roads connecting the villages, and the roads in Gower, once extremely poor, are now much improved.

Glamorganshire to some degree owes the comparative backwardness of its roads to the rapid extension of its railway system. The country is admirably served in this respect. A perfect network of lines covers the face of the land, and the remotest valleys have been rendered accessible by the engineer. With the exception of the Metropolis, Glamorganshire has more railways to the square mile than any other part of the kingdom. The railway naturally developed out of the tramroad, and Glamorganshire can claim the honour of having first discovered the possibilities of the locomotive. In 1802 a tramway was laid down from Merthyr to Abercynon for the conveyance of the product of the ironworks to the Glamorganshire Canal, which had previously been constructed. Over this track an experimental journey was made by Trevithick's high-pressure locomotive in 1804. The first regular railway constructed in Glamorgan was the Taff Vale, which commenced running between Merthyr and Cardiff in 1841.

Glamorganshire is now chiefly served by the Great Western, which absorbed the old South Wales Railway, and runs across the lowlands from Cardiff to Loughor. It has a number of local branches, and by means of the Vale of Neath line works also round the northern border of the county. Two other trunk lines, the London and North Western and the Midland, have connections between Swansea and the north ; the former also reaches Merthyr from Abergavenny. A great number of local lines bring the coal traffic from the hills to the coast. The Taff Vale and Rhymney railways link Cardiff with the eastern coalfield. The Barry railway serves the Barry docks, and the Port Talbot and the Rhondda and Swansea Bay lines are the channels by which the coal from the centre of the county reaches Swansea. The latter line penetrates the hills by a remarkable tunnel, two miles in length. A picturesque viaduct carries the Vale of Glamorgan railway across the glade of Porthkerry Park to Bridgend, a long trestle bridge takes the Great Western line over the Loughor river into Carmarthenshire, and a fine iron-girdered viaduct spanning the Rhymney valley enables the Barry railway to reach the Monmouthshire coalfield.

Though railways are the most efficient means of transport they were not the earliest. The age of railroads was preceded by an era of canals. Rivers are the natural channels of communication, but as none of the local rivers are navigable except at their estuaries, artificial waterways had to be constructed before the mineral wealth of the hills could reach the towns. The first canal to be made was the Glamorganshire Canal, which linked together Merthyr and Cardiff. Begun in 1790, it was opened for traffic in 1794, but the entire canal was not completed

Porthkerry Viaduct

till 1798. The difficulty of the fall of 600 feet between the two places had to be overcome by means of locks. In 1818 the canal was taken up the Cynon valley to Aberdare, and was afterwards extended from Cardiff to the sea. An almost contemporaneous undertaking was the construction of the Vale of Neath Canal from Glyn Neath to Aberdulais, the course of which was subsequently lengthened as far as Briton Ferry. Thirty years later the Tennant Canal was cut from Port Tennant, near Swansea, along the edge of the Crymlyn bog to Aberdulais in the Neath valley, and was carried across the Nedd by an aqueduct. Another engineering work executed about the same time was the Swansea Canal, which established communication between Swansea and the rich mineral district lying to the north. It runs up the Tawe valley to Ynysbydafau in Brecon, a distance of 17 miles, and though not as long as the Glamorgan Canal, the difficulties of construction were even greater, for it rises 873 feet. Though the canals in Glamorganshire, as elsewhere, were soon superseded by the railways, they have nowhere been worked with greater profit or enjoyed a longer period of prosperity. The traffic on them has now largely diminished, and some are obsolete. It has been said that the Act which authorised their construction was the Magna Charta of the county's prosperity.

23. Administration.

The political institutions of the Welsh were to a large extent determined by the physical peculiarities of the country which they inhabited. A land of mountains is seldom favourable to the social cohesion of its people. Wales never possessed any political unity. It was divided up into a number of petty kingdoms which were constantly at war one with another. The most that even the statecraft of the great Llewelyn could accomplish was to gather these principalities together into a sort of loose federation. Originally each of these Welsh kingdoms probably represented the "gwlad" or tribal area, distinguished one from another by differences of dialect and custom. Each kingdom consisted of a number of "cantrefs," which were the regions occupied by particular clans. Their limits corresponded in some degree with geographical distinctions. The kingdom of Morganwg was said to consist of seven cantrefs—Gorfynydd, which stretched from the mouth of the Tawe to the estuary of the Daw; Penychen, which reached from the Daw to the Taff; Y Breiniol, which lay between the Taff and the Rhymney; Gwynllwg, which was the territory enclosed between the Rhymney and the Usk; Gwent Iscoed and Gwent Uchcoed, which lay between the Usk and Wye; and Ewias and Erging, which were beyond the Monnow on the outlying borders of Herefordshire. The cantrefs displayed considerable independence, and had their own courts, but with the inevitable Welsh tendency to disintegration, as tribal ties became weaker, they eventually split up into smaller divisions, called "commotes," which were possibly assemblies of families, and which, like the cantrefs themselves, were ruled by their own customs and decided their own quarrels without much reference to any superior authority. Gower or Gwyr was a commote of one of the cantrefs of the kingdom of Deheubarth, and included not only the cliff-bound peninsula but the wilder country to the north between the Tawe and the Loughor. Among the Welsh, blood relationship was the one great bond, and the basis of the social organisation was the family. The political system was patriarchal, not territorial, and was built up on a foundation of kinship. The people were spread over the country in occupation of their own personal holdings and not, as in England, collected together in villages. To attain political union amid such conditions, except under the pressure of external necessity, was an impossibility, and political exigencies led eventually to the fusion of allied cantrefs into a small kingdom under the rule of the tribal head.

At the Norman conquest, so far as it effectually prevailed, the old Celtic system disappeared. The kingdom became the Norman lordship, and the lands were parcelled out to the different tenants on strict feudal tenure. The government was autocratic and arbitrary. Each fief, and each of the manors into which the fief was divided, had its own petty courts, but they severally exercised their authority under the paramount control of the county court of the lord, which recognised only the superior jurisdiction of the crown. The maintenance of the peace was a matter of local rather than of national concern, and the administration of justice was summary. The physical characteristics of the country and the independent spirit of the natives, however, compelled the Normans to modify to some extent their accustomed system. The more inaccessible regions were left in possession of the native chieftains, and were very largely allowed to retain their own laws and customs, though the overlord sometimes found it necessary to assert his authority by the expulsion of a local ruler. In Glamorganshire the districts of Senghenydd, Meisgyn, Glyn Rhondda, and Avon were under native rulers, and even the lordship of

Coity, though exercised by a Norman, is said to have retained to some extent its local customs. There is little doubt, however, that as time went on these native chieftains gradually assimilated their methods of government to the Norman system.

At the final incorporation of Wales with the English dominions in the reign of Henry VIII, the whole administration of justice passed directly into the hands of the crown. The lordship became a true shire, and was henceforth entitled to be represented in Parliament. The King's judges came on circuit to administer justice at the Great Session or Assize, and a number of local magistrates were appointed to deal with smaller offences at the petty sessions. Lords Lieutenant were appointed as the King's representatives, and the Sheriff became the King's executive officer for the enforcement of the law within the confines of the shire. For the better preservation of the peace and the punishment of the evil doer the shire was divided into a number of hundreds under the supervision of the Sheriff's bailiff. For legislative purposes each county and the corporate towns were accorded a representative in Parliament. Glamorganshire sent two members; one for the county, and one for the combined boroughs of Cardiff, Cowbridge, Llantrissant, Kenfig, Aberavon, Neath, Swansea, and Loughor.

In the reign of Elizabeth—in consequence of the increase of pauperism, due in part to the suppression of the monasteries, which had up to their dissolution acted as the voluntary almoners of the poor—the old parochial system was made a feature of civil government. Each parish became responsible for the maintenance of its own poor, and the affairs of the parish were managed by the Vestry. For the convenience of administration, however, the parishes have since been grouped into Unions.

This system of government has lasted practically into modern times. The Reform Act of 1832 and the Redistribution Act of 1885 introduced changes in the

Cardiff Town Hall and Law Courts

Parliamentary arrangements, and in 1889 a great extension of local government took place. The whole system of county administration was remodelled. County Councils were established for the general government of the shire, and the larger towns were made into county boroughs. The smaller corporate towns were allowed to retain their municipal powers, and localities which possessed no

W. G.   11

municipalities were placed under the control of Urban Councils, or grouped together under the authority of Rural Councils, according to the character of the population. The management of parish matters, which had hitherto been in the hands of the vestry, was transferred to a parish council, and the administration of the Poor Law was left, as before, to the existing Unions.

There are now within the county 3 county boroughs, 3 municipal boroughs, 15 urban and several rural district councils, and, for the administration of the Poor Law, 9 Poor Law Unions. The county boroughs are Cardiff, Swansea, and Merthyr Tydfil; the municipal boroughs are Aberavon, Cowbridge, and Neath. There are 126 civil parishes, and they are grouped together in Hundreds. The County Council consists of 88 members in all, 22 aldermen and 66 councillors. The aldermen sit for six years and are elected by the council, the councillors sit for three years and are chosen by the electors.

For the administration of justice, the county is placed under the jurisdiction of a Lord Lieutenant, a High Sheriff, and about 350 magistrates. It is included in the South Wales Circuit, and the Assizes are held alternately at Cardiff and Swansea.

Glamorganshire has now ten Parliamentary representatives. The county itself is divided into five districts —Mid, Southern, Eastern, Rhondda, and Gower—each of which returns one member. The combined boroughs of Cardiff, Cowbridge, and Llantrissant; the town of Swansea; and the district of Swansea, likewise return each a member, and the Merthyr district returns two.

Ecclesiastically the eastern half of the county is in the diocese of Llandaff, and contains the Cathedral city; while Swansea and Gower belong to the diocese of St David's.

The educational progress of Glamorganshire has been as remarkable as its industrial development. Before the

Cowbridge Grammar School

passing of the Welsh Intermediate Education Act of 1889 there were only four secondary schools in the shire, and, during the century preceding, practically the only place of regular instruction in the county at all was the ancient Grammar School of Cowbridge. Such education as existed was given by means of Sunday Schools and the so-called "Circulating Schools," which were bands of itinerating schoolmasters who visited the farmsteads to give occasional instruction to the inmates. These educational deficiencies were to some extent corrected at the beginning of the nineteenth century by the establishment in the industrial districts of "workshop" schools in connection with the larger works. The rural districts

University College, Cardiff

had to content themselves with such learning as could be obtained at the dame's school. The passing of the Education Act of 1870 made school attendance compulsory for the young, and brought into existence the School Boards. In 1889 special education committees were appointed to look after the interests of secondary education; but by the subsequent Act of 1902 the whole of the intermediate and elementary education of the county was placed in the hands of the County and Borough Councils, who have special committees to deal with the subject. There are now 20 intermediate and over 400 elementary schools in the county. The University College of Cardiff was opened in 1883, and has since become one of the constituent Colleges of the University of Wales. In addition to this centre of higher education there is at Swansea a Technical School, and a college for training female elementary teachers.

24. The Roll of Honour.

Welshmen have never lacked ability or enterprise, and many have attained to considerable distinction. Their fervent and imaginative temperament has frequently been wedded to more practical qualities, and men of affairs as well as poets and preachers have sprung from Welsh soil. Glamorganshire, however, has contributed comparatively few names of first-rate eminence to the roll of Welsh celebrities, though it has produced many whose achievements have been sufficiently notable to rescue their memories from oblivion. It is only possible here to give the names of the most remarkable.

Notwithstanding its stormy history Glamorganshire has had few soldiers of renown. All the prominent leaders were either foreigners or Welshmen from other counties. Of the local heroes the most celebrated was Ivor Bach, the great little lord of Senghenydd, whose exploit in scaling the walls of Cardiff Castle is still recounted with pride. From the same district came Llewelyn Bren, whose insurrection for a time disturbed the complacency of the English barons. Another man of the mountains whose fame still lives amongst his native haunts of the Rhondda was Cadwgan-y-Fwyell, the trusty henchman of Glyndwr. Colonel Philip Jones, the friend of Cromwell, lived at Fonmon, and Cromwell himself had distant connections with the county, for his great-grandfather, whose original name was Williams, was a native of Llanishen. In later times Sir William Nott (1782–1845), the son of a Neath publican, distinguished himself by his successful military operations at Candahar.

In spite of its maritime character Glamorganshire has reared few sailors of eminence. The most prominent were Admiral Sir Thomas Button (d. 1634), a native of Cardiff who explored Hudson's Bay, and did good service by suppressing the Channel pirates ; and Admiral Mathews of Llandaff (1670–1751), whose promising career was ruined by the miscarriage of his expedition against the French at Toulon.

In religious devotion the Welsh have always been conspicuous. A number of noteworthy ecclesiastics and divines have been connected with Glamorganshire's religious history, though they have not all been sons of the soil. The association of the names of Teilo and Dyfrig with the see of Llandaff is probably fanciful. Cadoc, the reputed founder of Llancarfan, on the contrary, undoubtedly exercised considerable influence in the locality, though he sprang from the princely house of Gwynllwg.

Another equally famous name in the annals of local monasticism is that of Illtyd, the head of Llantwit Major. Of the bishops of Llandaff the most famous were Urban

Admiral Sir Thomas Button

(c. 1107), the builder of the first Norman cathedral, who is reputed to have been of local birth ; William Morgan, the translator of the Welsh Bible, who occupied the see

from 1595 to 1601 and was the first Welshman to preside
over the diocese after a lapse of three centuries; and
Morgan Owen, Laud's chaplain.   Of the many Nonconformist divines to whom the county has given birth
the most eminent were Christopher Love (1618–1651),
a Cardiff Presbyterian, who acquired notoriety for his
anti-monarchical opinions, but was eventually put to death
as a Royalist; Thomas Llewelyn of Gelligaer (d. 1783),
a Biblical scholar; Dr Richard Price, the philosopher,
1723–1791; and T. W. Davies, a native of Gower, a
Nonconformist historian of repute.   Christmas Evans
(1766–1838), though not a Glamorganshire man, attained
wide celebrity for his pastoral labours at Caerphilly
and Cardiff.   The most distinguished local Papist was
Sir E. Carne of Cowbridge (d. 1561), who acted as
ambassador at Rome for both Mary and Elizabeth.

Several notable men of letters have had associations
with the county though they have not all been Welshmen.
To Caradoc of Llancarfan (d. 1157) we owe some of
the legendary matter which has built up the Arthurian
romances.   John Walters (d. 1794), rector of Llandough
near Cowbridge, was an eminent Welsh lexicographer.
John Sterling, whose life Carlyle wrote, spent his early
days at Llanbleddian; Lady Charlotte Guest, the translator of the *Mabinogion*, was the wife of Sir John Guest,
the Merthyr ironmaster; the voluminous Ann Kemble,
sister of Mrs Siddons and popularly known as "Ann
of Swansea," passed most of her life at Swansea;
R. D. Blackmore, the novelist, sojourned for a time at
Newton Nottage; and Thomas Bowdler, the notorious

expurgator of Shakespeare, is buried at Oystermouth. The local poets have been more numerous than distinguished. The birthplace of Dafydd ap Gwilym (1330–68) is doubtful,

Beau Nash

but in no part of Wales was he held in higher honour than in Glamorgan. Swansea claims, though on insufficient evidence, to have given birth to John Gower (d. 1402) the personal friend of Chaucer. Amongst less notable songsters were Meirig Dafydd, Dafydd Benwyn (*c.* 1560), Rhys Brydydd, Llewelyn Sion, Lewis Morganwg (1500–1540) and Thomas Leyson of Neath (1569–1607). Glamorganshire's most notable musicians were W. T. Rees (b. 1838), a Bridgend collier, and Joseph L. Parry of Merthyr, Professor at Cardiff College (1841–1903).

There are few scholars of more than local celebrity. Sir Edward Stradling of St Donat's (1529–1609) was a generous patron of learning and renowned as a collector of manuscripts, and other members of the same family won repute for scholarship. One of the earliest pioneers of Welsh education was Sir Leoline Jenkins of Llantrissant (1625–85), the second founder of Jesus College, Oxford ; and another statesman who helped to complete the educational system of the Principality was the first Lord Aberdare (1815–1893), the first vice-chancellor of the University of Wales, who was born at St Nicholas.

Two Glamorganshire men of humble parentage won considerable local reputations as antiquaries. One was Edward Williams, better known as "Iolo Morganwg" (1743–1826), a mason of Penon, and the other Evan Davies (1816–84), a watchmaker of Pontypridd. An antiquary of wider fame was George Clark (1809–98) who, though not a native, spent a large part of his life as manager of the Dowlais ironworks and wrote on archaeological subjects.

Of legal celebrities the most notable occupants of the Bench were Judge David Jenkins of Hensol (1582–1663), whose learning procured for him the sobriquet of "the

John Crichton Stuart, 2nd Marquis of Bute

Pillar of the Law"; Sir John Nichol of Llanmaes (1751–1838), who was a judge of the Admiralty and Vicar General; and Sir William Grove (1811–1896), born at Swansea, who combined a Judgeship in the High Court of Justice with great distinction as a physicist and electrician, being Professor at the London Institution from 1840 to 1847.

Amongst local craftsmen Richard Twrch of Bridgend the designer of the renaissance porch at Beaupre, and William Edwards (c. 1755) the builder of the single-arch bridge at Pontypridd, should be remembered. Glamorganshire has contributed one notable figure to the world of fashion in Richard Nash, "Beau Nash" (1674–1762), the famous M.C. of the Bath routs, who was a Swansea man.

Of the great captains of industry whose enterprise revolutionised the commercial prospects of the county the most conspicuous were Dr Lane, the founder of the copper-smelting trade; Sir J. Guest, Richard Crawshay, and Richard Fothergill, the ironmasters; and John Crichton Stuart, 2nd Marquis of Bute, the creator of modern Cardiff.

25. THE CHIEF TOWNS AND VILLAGES OF GLAMORGANSHIRE.

(The figures in brackets after each name give the population or the place in 1911, from the official returns, and those at the end of each paragraph are references to the pages in the text.)

Aberavon (10,505), a municipal and market town at the mouth of the Avon. The town, wedged in between the

Aberthaw Village

mountains and Swansea Bay, possesses a promenade and some extensive sands. Its corporate charter was obtained as early as 1158 A.D., and it once enjoyed Parliamentary privileges.

Collieries and tin-plate works are the chief sources of employ-
ment. (pp. 34, 51, 160, 162.)

Aberdare (50,830), a market town in the Cynon valley. It
is a place of modern growth and owed its early prosperity to the
Abernant ironworks. It now depends chiefly on its collieries.
(pp. 28, 30, 87, 90.)

Aberthaw, a village and decayed port at the mouth of the
Daw. It was once famous for its lime, which was used by
Smeaton in the construction of the Eddystone Lighthouse. (pp.
37, 49, 86.)

Barry (33,763), a large sea-port in the south of the county.
Its career has been one of extraordinary prosperity. Fifty years
ago it did not exist; to-day it is one of the most promising ports
in South Wales, and its docks are some of the finest in Glamorgan.
The town forms an amphitheatre round the docks, and comprises
the districts of Cadoxton, Barry Dock, Barry, and Barry Island.
The last, now an island only in name, is regarded as the local
holiday quarter, and possesses a bathing place. On the summit
of the cliff are the foundations of an ancient chapel, and on the
mainland is the gateway of a medieval castle. (pp. 25, 37, 48, 102.)

Bishopston (893), a village in the south of the Gower
peninsula. A charming glen runs down from the village to the
sea. Overhanging the mouth of the valley is Pwlldu Head, a fine
limestone promontory crowned by a camp. (p. 52.)

Boverton, a hamlet near Llantwit Major, believed by some
to have been the *Bovium* of the Romans. Near it are the ruins
of a fortified manor house, Boverton Place. (p. 128.)

Bridgend (8021), a market town on the Ogmore river,
which divides it into two quarters, Old Castle and New Castle.
Each township formerly possessed a fortress, but the ruins of
New Castle, which retains a Norman gateway, alone remain.
(pp. 20, 25, 86, 93, 148.)

Briton Ferry (8472), a sea-port at the mouth of the Nedd, deriving its importance from its docks, and steel and tin-plate works. (pp. 35, 51, 82, 103.)

Caerau (237), a village four miles west of Cardiff. The church stands on an eminence, which is surmounted by a large Roman camp. (p. 130.)

Caerphilly (32,844), a market town on the east border of the county, seven miles north of Cardiff. The town, which is

Bridgend

pleasantly situated at the foot of Cefn Carnau, owes its large population to the neighbouring colliery district of the Rhymney valley, though even in the seventeenth century its market rivalled that of Cardiff. It is chiefly notable for the ruins of its famous castle, built by Gilbert de Clare in the thirteenth century, and enlarged by the Despensers. In size it is second only to Windsor, and possesses a fine hall and some very elaborate outworks. In the vicinity are one or two old manor houses. (pp. 20, 30, 111, 145.)

Cardiff (182,259), a county and parliamentary borough on the Taff, and the principal town in Glamorgan. Though Cardiff has been termed the Welsh Chicago, it is a place of considerable antiquity. As *Tibia Amnis* it was a military station of importance in Roman days, and one of its gateways has been discovered in the grounds of the castle, which was built on the lines of the Roman rampart. The castle itself is an elaborate modern restoration of the medieval fortress, but the ruins of the Norman keep built by Robert of Caen stand on a moated mound in the centre of the court. The Curthose tower is said to have been for 20 years the prison of Robert of Normandy. Besides the castle the only other ancient building in the town is St John's church, a fifteenth century edifice with a fine Perpendicular tower. The old church of St Mary, which was connected with a Benedictine priory, was destroyed by a flood in 1607. Cardiff also once possessed some habitations of Black and Grey Friars, and the ruins of the house which Lord Herbert built out of the material of the latter still stand near the City Hall. Though always a port, Cardiff's commercial prosperity dates only from the middle of the last century. It now possesses a splendidly equipped series of docks, and it is the largest coal port in the world. The town is well built and its streets are spacious. A fine group of buildings consisting of the City Hall, the Assize Courts, the Welsh National Museum, and the University College have been erected in Cathays Park, and there is a large library in another part of the town. (pp. 3, 27, 31, 46, 80, 82, 95, 98, 99, 101, 109, 111, 113, 115, 116, 128, 138, 144, 145, 160, 165.)

Castell Coch, a castle on the slopes of Cefn On, three miles south-west of Caerphilly, which takes its name from the ruddy tint of its walls. The present structure is a modern restoration of the original building, which was probably intended to guard the pass of the Taff. On the hill side are Lord Bute's well-known vineyards. (pp. 77, 144.)

Cheriton (128), a village in the north-west of the Gower peninsula. The church is Early English, and at Llandimore are some remains of a fifteenth century castle. On North Hill Tor is a Danish camp. (pp. 55, 138.)

Coity, a village near Bridgend. It possesses a fine cruciform Decorated church, and the crumbling remains of a large castle, once the seat of the Turbervilles. (pp. 138, 145.)

Cowbridge (1167), a municipal borough and market town pleasantly situated in the centre of the Vale of Glamorgan on the

Cheriton Church

banks of the Daw. It shares in the Parliamentary representation of Cardiff. The church has a curious Early English tower, and near it are some fragments of the ancient town walls and one of the gateways. The Grammar School was refounded in the time of James II. The neighbouring hill of Stalling Down was the scene of a sanguinary encounter between Glyndwr and the English forces in 1405. (pp. 25, 36, 114, 141, 145, 160, 162, 163.)

Coychurch (1391), a village near Bridgend possessing a handsome cruciform church of the Transitional Decorated period. In the churchyard are two Celtic crosses. (pp. 23, 132, 138.)

Dinas Powis, a village five miles south-west of Cardiff, where there are the remains of a castle.

Ewenny (357), a parish on the Ewenny river near Bridgend, notable as the site of a Benedictine priory founded in 1140 by Maurice de Londres. The church remains almost intact and is a fine example of Norman work. The gateway and portions of the external wall of the monastery also survive. (pp. 33, 136.)

Flemingston (69), a village 3½ miles south-east of Cowbridge overlooking the Daw. Near the church is a once fortified manor house with a Jacobean hall. (p. 149.)

Gelligaer (35,521), a village on the east border of the county 14 miles north of Cardiff overlooking the Rhymney valley. It was once a Roman military station, and considerable remains of the fortifications have recently been unearthed. In the neighbourhood are the populous colliery villages of Brithdir, Bargoed, Hengoed, and Pontlottyn. (pp. 128, 130, 168.)

Glyncorrwg (8688), a colliery district amongst the hills, six miles north of Maesteg. (p. 69.)

Gorseinon, a village near Loughor depending on vitriol and tin-plate works. The ancient parish church of Llandeilo Talybont is interesting. At the neighbouring village of Gowerton are some extensive steel works.

Hirwain, a populous colliery suburb of Aberdare, four miles north-west of the town. (pp. 9, 92.)

Kenfig (301), a village amongst the sand-warrens near Porthcawl. It was once a corporate and parliamentary borough, but the town was overwhelmed by the sand in the sixteenth century. A fragment of its castle may still be seen amongst the dunes, and the local public house was once the town hall.

The old municipal mace is in the Cardiff Museum.   In the centre of surrounding sand-hills is Kenfig Pool, a sheet of water two miles round, and on the road to Margam is an Ogam stone. (pp. 42, 44, 51, 59, 91, 109, 126, 145, 160.)

Laleston (706), a village near Bridgend where are the remains of a cromlech.

Llanbleddian (745), the mother parish of Cowbridge.   On a knoll overlooking the village are the remains of St Quintin's castle.   (pp. 114, 168.)

Llancarfan (446), a village in the valley of the Carfan brook, a feeder of the Daw.   It was the site of a famous Celtic monastery founded by St Cadoc in the sixth century.   The church has a transitional Norman arcade, and the remains of a fine screen. On a neighbouring hill is an ancient camp.   (pp. 132, 138, 166, 168.)

Llandaff (9142), a cathedral city on the banks of the Taff, near Cardiff.   It was a bishopric of importance before the coming of the Normans, but the present cathedral dates from the twelfth century, and though retaining its Norman doorways, belongs chiefly to the Early English and Decorated periods.   It possesses a number of interesting monuments.   On the high ground above the cathedral are the ruins of the ancient episcopal palace destroyed by Glyndwr in 1402.   (pp. 23, 32, 113, 134–6, 150, 167.)

Llanddewi (120), a parish in the west of Gower in the vicinity of which are the sites of Llanddewi and Scurlage castles. (p. 144.)

Llandough, a village near Penarth, once the site of a Celtic monastery, and still retaining a fine specimen of a Celtic pillar cross.   There is another village of the same name near Cowbridge.   (pp. 25, 93, 132.)

Llangan (207), a small village 3½ miles north-west from Cowbridge possessing an interesting example of a sculptured

12—2

wheel cross of Celtic workmanship, and an almost perfect fifteenth century cross. (pp. 132, 138.)

Llangenydd, a village on the west verge of Gower, once the site of a monastery. On Harding Down are some earthworks. (p. 55.)

Llangyfelach, a parish five miles north of Swansea. On one of the neighbouring hills is a stone circle, and in the churchyard is the base of a Celtic cross. (p. 125.)

Llangynwyd (2098), the mother parish of Maesteg, near which are the remains of a castle destroyed by the younger Llewelyn. (pp. 110, 144.)

Llanmadoc (149), a village on the north-west extremity of the Gower peninsula. In the churchyard is a Celtic cross, and on the summit of the neighbouring hill is a large camp. The shore is overhung in places by some prominent limestone tors beneath which are bone caverns. (p. 55.)

Llanmaes (142), a village four miles south from Cowbridge. The church contains some mural frescoes. Near it are the ruins of Malafant Castle. (pp. 114, 141.)

Llanmihangel (30), a parish near Cowbridge possessing a fine fifteenth century manor house remodelled in Tudor times. (p. 148.)

Llanrhidian, a village in Gower bordering on the Burry inlet. In the vicinity are the ruins of Weobley Castle, and on the ridge of Cefn-y-Bryn is the famous cromlech, Arthur's Stone. The parish also possesses more than one menhir and some traces of a camp. (pp. 55, 58, 124, 145.)

Llansamlet (3801), a parish in the Swansea industrial zone overlooking the Tawe valley, thriving chiefly on tin-plate and spelter works. (p. 82.)

Llantrissant (50,929), a market town on the north edge of the Vale of Glamorgan, and a contributory borough to Cardiff. It is chiefly remarkable for its romantic situation on the brow of a hill overlooking the lowlands. The eminence is crowned by the ruins of a castle said to have been destroyed in the insurrection of Llewelyn Bren. On the opposite hill is a camp. Iron ore was once worked in the neighbourhood, which is now occupied by colliery villages. (pp. 20, 32, 80, 83, 113, 160, 162, 169.)

Llantrithyd, a village three miles eastward from Cowbridge, possessing an interesting church and the ruins of a Tudor manor house. (p. 150.)

Llantwit Major (1188), a quaint old town 4½ miles south-west of Cowbridge on the Colhugh brook, and to the antiquary one of the most attractive places in the county. In the sixth century it was the seat of a monastery presided over by St Illtyd, though nothing remains from Celtic times but a collection of ninth century crosses in the church, which dates from the thirteenth century and is of peculiar interest. The town possesses a medieval town hall, and the ruins of a manor house. In a neighbouring field were discovered the foundations of a Roman villa. Overlooking the shore are some earthworks. (pp. 37, 49, 130, 131, 137, 138, 148, 167.)

Loughor (4118), a decayed town at the mouth of the Loughor river eight miles north-west of Swansea, generally identified with the Roman *Leucarum*. At the vicarage is an Ogam stone made out of a Roman altar, and by the side of the river are the ruins of a medieval castle. (pp. 7, 36, 126, 128.)

Maesteg (24,977), a flourishing town in the Llynfi valley some six miles eastward of Port Talbot, owing its existence to its former ironworks but now chiefly dependent upon its collieries, which stretch from the adjoining village of Garth northwards to Caerau. (p. 33.)

Marcross (95), a village near Nash Point, six miles south-west of Cowbridge. The church contains a Norman chancel arch, and in the parish are the remains of a cromlech. (pp. 49, 60, 138.)

Margam (14,713, including Port Talbot), a village four miles south-east from Aberavon, noteworthy for its once prosperous Cistercian monastery founded by Robert of Gloucester in 1147. The site of the abbey is now occupied by a mansion, in the grounds of which are the ruins of the choir and chapter-house. The parish church, a late Norman building, was the nave of the minster and contains a fine collection of Celtic crosses and sepulchral slabs. There are several inscribed stones and camps on the neighbouring hills. (pp. 14, 88, 128, 131, 136.)

Merthyr Mawr (189), a parish near Bridgend on the Ogmore river. In the churchyard are some effigies and Celtic crosses, and two other crosses are at Merthyr Mawr House. The shore is remarkable for its extensive sand-warrens, on the edge of which stand the remains of Candleston Castle. Among the sand-hills have been discovered remains of a large prehistoric burial-place. (pp. 87, 132.)

Merthyr Tydfil (80,990), a county and parliamentary borough in the north-east corner of the county, now the third most important town in the county and the metropolis of the iron trade. Adjoining it are the Dowlais ironworks, and the neighbouring collieries are very numerous. To the north of the town are the ruins of Morlais Castle, a feudal outpost. (pp. 3, 30, 81, 82, 86, 87, 89, 144, 162, 169.)

Monknash (78), a village on the Bristol Channel, near the Nash. It borrows its prefix from some remains of monastic buildings.

Mountain Ash (42,246), a typical colliery town in the Cynon valley 3½ miles south-east of Aberdare. (p. 30.)

Nantgarw, a village in the Taff valley once celebrated for its porcelain. (p. 86.)

Neath (17,586)—the *Nidum* of the Romans—a municipal borough and market town on the Nedd. In medieval times it possessed both a castle and an abbey, the ruins of which still remain. The abbey, described by Leland as the "fairest abbey in Wales," was founded for Cistercians by Richard de Granville in 1111, and was Early English in style. Portions of the church

Merthyr Tydfil

survive, but the site is chiefly occupied by the ruins of a seventeenth century mansion erected out of its materials and incorporating some of its domestic buildings. The town now is the centre of a busy industrial district abounding in tin-plate and copper works. On the summit of Mynydd Drummau are a stone circle and a menhir. (pp. 35, 82, 86, 125, 128, 137, 150, 162.)

Newton Nottage (3444, including Porthcawl), a village on Ogmore Bay near Porthcawl. The church has a fortified tower and a curious stone pulpit. Between the church and the

sea is a well which fills only when the tide is out. Nottage Court
is a Tudor manor house. (pp. 23, 51, 139, 149, 168.)

Ogmore Valleys, a thickly populated colliery district
through which flow the Ogmore and Garw rivers. The chief
villages are Nantymoel, Gilfach Coch, Blackmill, Blaengarw and
Pontycwmmer. (pp. 33, 90.)

Oxwich (178), a village in the south of Gower, on the
shores of a large and beautiful bay. The church is interesting,
and on the headland are the ruins of a castle. (pp. 52, 58.)

Oystermouth (6098), a seaside resort on the west side of
Swansea Bay, better known from its outstanding islets as the
Mumbles. The village possesses a pier and the picturesque ruins
of a castle. Beyond the headland are Langland and Caswell
Bays, two favourite bathing-places. (pp. 52, 60, 97, 145, 168.)

Penarth (15,488), a watering place and port on the Ely
estuary, two miles south of Cardiff. Its proximity to Cardiff and
its fine position on the top of a cliff have made it a favourite
residential quarter. It possesses some docks and a pier. A section
of Rhaetic rocks, from their extensive exposure on the headland,
are known as "Penarth Beds." (pp. 23, 46, 75, 86, 101.)

Penmark (495), a village in the south of the Vale of
Glamorgan three miles west of Barry, where are the ruins of a
castle. Fonmon Castle, a modernised Norman stronghold, is also
in the neighbourhood. (pp. 69, 144, 166.)

Pennard (245), a parish in Gower on the east side of Oxwich
Bay. Overlooking a sandy creek are the ruins of a castle, and
the face of the cliffs are perforated with bone caverns. The
neighbouring village of Parkmill is much visited for its
picturesqueness. (p. 54.)

Penrice (215), a village in Gower near Oxwich Bay possessing
the picturesque ruins of a once strong and extensive castle. (p. 54.)

Peterston-super-Ely (389), a village in the Vale of Glamorgan, six miles west of Cardiff, where are a few fragments of a castle. (p. 32.)

Pontardulais, a village in the Loughor valley, the centre of a busy district occupied chiefly in tin-plate manufacture.

Pontypridd (43,211), a large market and colliery town at the confluence of the Rhondda and Taff rivers forming a natural focus upon which the immense traffic from the contiguous mining valleys converges. It possesses a remarkable one-arched bridge, and on the hill-side overlooking the town is a modern stone circle. Hopkinstown and Treforest are industrial suburbs. (pp. 31, 85, 87, 169.)

Port Eynon (207), a village in the south of the Gower peninsula standing on its own bay, and possessing some fine cliff scenery. In the neighbourhood are the Culver Hole and the Paviland Caves. (p. 54.)

Porthcawl (3444, including Newton Nottage), a popular watering place on the west side of Ogmore Bay. It possesses a dock, but its shipping trade has now vanished. Near it is Sker House, a decayed Tudor manor house, associated with Blackmore's *Maid of Sker*. (pp. 51, 103, 149.)

Porthkerry (181), a parish near Barry overlooking a pretty bay. On the cliffs are some earthworks known as "the Bulwarks." (pp. 48, 155.)

Port Talbot (14,002, including Margam), a flourishing port on the east side of Swansea Bay, with some extensive docks and steel works. (pp. 34, 103.)

Rhondda Valleys (152,781), the most productive colliery district in the county, comprising the villages along the banks of the two Rhondda rivers and their tributaries. It is celebrated for its steam coal, and is densely populated. The chief places are

Blaen Rhondda, Treherbert, Treorky, Pentre, Ystradyfodwg, Llwynypia, Clydach, Tonypandy, Ferndale, Tylorstown, and Porth. On the hill side near Llwynypia once stood the Franciscan Friary of Penrhys. (p. 30.)

Rhossili (246), a village at the south-west extremity of Gower near the Worms Head. The cliff scenery is bold and impressive, and behind the bay rise extensive downs on which are some prehistoric antiquities. (pp. 55, 138.)

St Athan (360), a village near the mouth of the Daw six miles south-east of Cowbridge. The church has some interesting monuments, and on the banks of the river are the ruins of East Orchard castle.

St Bride Major (783), a village four miles south of Bridgend in a fold of Ogmore Down. The church has some monuments, and near the confluence of the Ewenny and Ogmore rivers is the ruined keep of Ogmore Castle. (pp. 33, 145.)

St Donat's a village one mile east of Nash Point possessing a residential castle of great interest. In the churchyard is a fine medieval cross. (pp. 49, 141, 144, 169.)

St Fagan's (549), a village four miles west of Cardiff. A desperate engagement between the Royalist and Parliamentary forces took place in the neighbourhood in 1648. St Fagan's Castle is a large Tudor residence built on the site of a medieval stronghold. The church exhibits some Decorated work. (pp. 32, 117, 139, 149.)

St Hilary (138), a village two miles south-east of Cowbridge overlooking the valley of the Daw. In the parish are the ruins of Beaupré Castle which possesses a fine renaissance porch. (p. 149.)

Swansea

St Mary Hill (177), a parish four miles east of Bridgend noted for its annual horse fair. The church is a prominent landmark and possesses a good medieval cross. (p. 114.)

St Nicholas (370), a village mid-way between Cardiff and Cowbridge. In the neighbourhood are some remarkable cromlechs. (p. 124.)

Southerndown, a watering place four miles south of Bridgend. The coast is geologically interesting. Behind a lofty headland is Dunraven Castle, a modern mansion occupying traditionally the site of an early castle. (p. 50.)

Sully (314), a village on the Bristol Channel two miles east of Barry. Near the church are the foundations of a castle and close to the coast is a small island on which is a camp. (pp. 48, 144.)

Swansea (114,663), a county and parliamentary borough at the mouth of the Tawe, and the second largest port and town in the county. It is supposed to have originated as a Scandinavian settlement, but its real history begins with the Norman castle built by the Earl of Warwick on his conquest of Gower. The ruins of this castle as reconstructed by Bishop Gower in 1330 stand in the centre of the town and show some Decorated arcading. The town itself occupies a fine situation on the shores of its bay, and the suburbs are spread over the adjoining hill side. In spite of its long history it has few antiquities. The chief public buildings are the Town Hall, Royal Institution, and Free Library. The church, though modern, has an ancient chapel attached. The local commerce is immense, and the town possesses some magnificent docks. It is the metropolis of the copper trade and owing to the number and variety of its smelting works it has been termed the metallurgical capital of the world. There are more than 150 furnaces for the treatment of different kinds of ores within four miles of the town. (pp. 3, 48, 52, 82, 85, 88, 98, 99, 103, 162, 168.)

Swansea Valley (111,950), the populous district lying to the north of Swansea. It consists of a long string of industrial villages following the course of the Tawe, the chief of which are Ystalyfera, Llanguick, Pontardawe, Clydach, Morriston, and Landore. The old Swansea canal runs through the valley, which abounds in collieries and copper, steel, and chemical works. The northern portion contains some valuable anthracite seams. At Ystalyfera there is some striking rock scenery, and on a hill near Llanguick are the remains of a stone circle. (pp. 36, 82, 83, 85, 135.)

Tondu, a market town at the junction of the Llynfi and Ogmore valleys, depending on collieries and ironworks. (p. 33.)

Wenvoe (505), a village 4½ miles south-west from Cardiff which once possessed two castles, Wenvoe and Wrinstone.

Ystradowen (228), a village two miles north of Cowbridge near which are some fragments of Talyfan Castle.

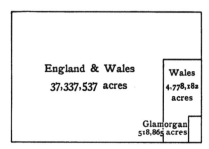

Fig. 1. The Area of Glamorganshire compared with
that of England and Wales

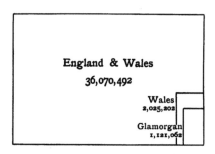

Fig. 2. The Population of Glamorganshire compared
with that of England and Wales

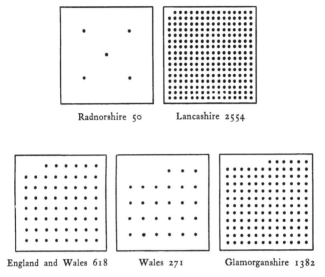

Fig. 3. Comparative Density of Population to the square mile in 1911

(Each square represents a sq. mile and each dot ten persons)

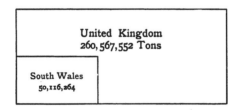

Fig. 4. Output of Coal from the S. Wales Coalfield in 1912 compared with that of the United Kingdom

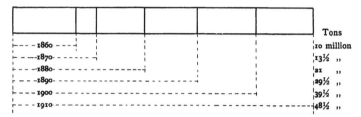

					Tons
1860					10 million
1870					13½ ,,
1880					21 ,,
1890					29½ ,,
1900					39½ ,,
1910					48½ ,,

Fig. 5. Comparative Increase in the Output of Coal from
the S. Wales Coalfield during the last 50 years

				Total Tonnage Imports & Exports
1863				4,035,827 Tons
1870				4,323,089 ,,
1880				8,273,029 ,,
1890				15,452,502 ,,
1900				21,895,445 ,,
1910				28,349,033 ,,

Fig. 6. Growth in Trade of the Glamorganshire Ports
during the last 50 years

1801					70,879
1851					231,849
1861					317,752
1871					397,859
1881					511,933
1891					687,218
1901					859,931
1911					1,121,062

Fig. 7. Growth in the Population of Glamorganshire
during the last century

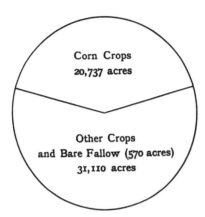

Fig. 8. Proportionate Area under Corn Crops
in Glamorganshire in 1912

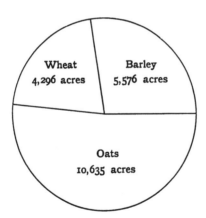

Fig. 9. Proportionate Areas of chief Cereals
in Glamorganshire in 1912

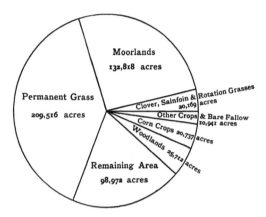

Fig. 10. Proportionate Areas of land in
Glamorganshire in 1912

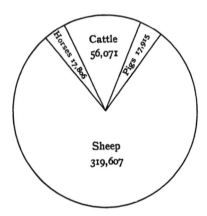

Fig. 11. Proportionate numbers of Live Stock
in Glamorganshire in 1912

Fig. 12. Value of Glamorganshire Imports and Exports
compared with that of the United Kingdom in 1912

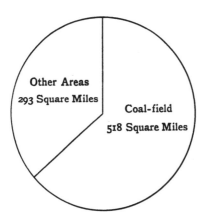

Fig. 13. Proportionate Areas of Coal-producing
and other districts in Glamorganshire

Fig. 14. Comparative Value of Imports and Exports at the various Glamorganshire ports in 1912

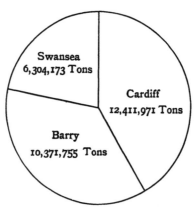

Fig. 15. Comparative Volume of Trade of the chief Glamorganshire ports in 1912

www.ingramcontent.com/pod-product-compliance
Ingram Content Group UK Ltd.
Pitfield, Milton Keynes, MK11 3LW, UK
UKHW042143280225
455719UK00001B/60

9 781107 619722